— TRACKING —
ROCKY MOUNTAIN
RAILROAD CLUB
EXCURSIONS
— 1987–1990 —

MIKE BUTLER

America Through Time is an imprint of Fonthill Media LLC
www.through-time.com
office@through-time.com

Published by Arcadia Publishing by arrangement with Fonthill Media LLC
For all general information, please contact Arcadia Publishing:
Telephone: 843-853-2070
Fax: 843-853-0044
E-mail: sales@arcadiapublishing.com
For customer service and orders:
Toll-Free 1-888-313-2665

www.arcadiapublishing.com

First published 2023

Copyright © Mike Butler 2023

ISBN 978-1-63499-450-7

Typeset in 10pt on13pt Sabon
Printed and bound in England

Acknowledgments

The late 1980s saw a flurry of activity from the Rocky Mountain Railroad Club as the club prepared to celebrate its fiftieth anniversary in 1988. Founded in 1938 in Denver, the club's golden anniversary was to be celebrated with rail excursions, banquets, publications, and commemorative items such as a belt buckle, calendar, and HO-scale model caboose. This incredible celebration was thanks to the many volunteers who put it all together and the leadership of Club President John Dillavou. This book would not have been possible without all of those volunteer efforts and I am grateful to one and all. Special thanks to my good friend and co-worker at Denver Parks and Recreation, John Dillavou. Many stressful work moments were relieved as we gazed from our office window at all the train activity passing by below us on the Central Platte River Valley rail corridor. Thanks also go to my wife, Mary Jane, and son, Jason, who supported all my journeys with the club and joined in on some of the excursions.

When I first conceived this book, I thought it would just be about the Rocky Mountain Railroad Club's golden anniversary excursions of 1988. However, looking back over my journals and photographs, I realized I had only been able to participate in one of the excursions in 1988. Thus, the book was expanded to include the excursions I participated in and photographed from 1987–1990. They were very representative of the club's activities in the golden anniversary era. I hope readers enjoy the stories and photos from those excursions.

All the photographs in the book were taken by the author except for those otherwise noted in the photo captions. Most of the author's photos were Kodachrome slides, which are now approaching thirty-five years old, so apologies for some darkness and fading. While researching this book, the author consulted many sources noted in the bibliography, however any errors in the text are strictly his own.

Contents

1

The Silver Sky to Winter Park: March 28, 1987

Departing Denver Union Station at 7:30 a.m. on Saturday March 28, 1987, the Rocky Mountain Railroad Club's chartered car "Silver Sky," trailing at the end of the Denver & Rio Grande Ski Train, headed 57 miles across the Rocky Mountains to Winter Park. The Silver Sky was a sleek silver-sided dome/observation car in contrast to the gold, silver, and black passenger cars of the Ski Train. Those cars were in fact leftovers from the Northern Pacific Railroad and, at seventy-nine years of age, were approaching the end of their useful life. The Silver Sky had been built by the Budd Manufacturing Company in 1948 and was part of the California Zephyr consist, which began service in 1949 from Chicago to Oakland, California.

As the Silver Sky began its journey west at the end of the Ski Train, it climbed into the Rocky Mountain foothills west of Denver on a series of S-curves. It soon entered the "Tunnel District," which originally consisted of thirty tunnels through the mountains. The finale was the 6.21-mile-long Moffat Tunnel, which disgorged the train on the western slope of the Continental Divide at the base of the Winter Park Ski Area. The Rio Grande Ski Train had been running since 1947, though there had been earlier "snow trains" to Winter Park and Hot Sulphur Springs.

None of this activity would have been possible were it not for the dream of David H. Moffat to build a railroad from Denver to Salt Lake City. Moffat arrived in Denver in 1860 at the age of twenty-one. He started a "stationery" store, which not only sold books and paper, but all kinds of supplies to miners and travelers. He was an acute businessman and he invested in the mines west of Denver. His financial success gave him the resources needed to eventually build a railroad and he caught the attention of William Jackson Palmer, founder of the Denver & Rio Grande Railroad. Palmer, ready to retire, asked Moffat to become president of the D&RG in 1885. Moffat accepted the position and immediately sent surveyors west to find a shorter route across the mountains than the lengthy D&RG route south to Pueblo and then back northwest to Leadville. When Moffat was fired by the D&RG in 1891, he continued his search for a direct rail route west from Denver. This led to his formation of the Denver,

The Silver Sky awaits its passengers at Denver's Union Station, March 28, 1987.

Rocky Mountain Railroad Club President John Dillavou signed the tickets for the excursion.

The Ski Train heads into one of the tunnels on its way west to Winter Park on a snowy March day.

The Silver Sky rests at the West Portal of the Moffat Tunnel (seen at the lower right) after its journey from Denver.

Northwestern and Pacific Railroad in 1902 and the construction of some thirty-three tunnels through the mountains west of Denver over Rollins Pass:

> Moffat's reasons for building the Denver, Northwestern and Pacific Railroad were to put Denver on a direct transcontinental rail route to Salt Lake City and the Pacific Coast and to develop the natural resources of northwestern Colorado, thus increasing Denver's business position in the region. The proposed route would be just about midway between the Union Pacific to the north and the Denver & Rio Grande to the south and would serve a vast area that at present had not one mile of railroad.[1]

Construction of the Denver, Northwestern and Pacific Railroad continued into the summer of 1904:

> On June 23, 1904, the first train of the Denver, Northwestern and Pacific steamed west from Denver up almost continual two percent grades, through seemingly endless tunnels, to open one of the most colorful chapters in American railroad history.[2]

Rails proceeded around Yankee Doodle Lake and then through the 170-foot Needle's Eye Tunnel until reaching Rollins Pass at 11,660 feet elevation. "On September 2, 1904, locomotive no. 300 pulled three passenger coaches to the 'Top of the World.'"[3] The station at the top of the pass was called "Corona," and construction continued west from there.

On April 30, 1913, the Denver, Northwestern & Pacific Railroad was reorganized as the Denver and Salt Lake Railroad. The extreme elevation of Rollins Pass caused continual hardships for the railroad. Winters were brutal. Snow closed the line for days on end. Trains were stranded. This problem was eventually solved by the construction of the Moffat Tunnel under the Continental Divide from 1923–1927. Unfortunately, David Moffat died in March 1911, never witnessing the completion of his namesake tunnel:

> The new route via the tunnel was 22.8 miles shorter than over Rollins Pass and 2,421 feet lower in … altitude … The tunnel was officially opened on February 26, 1928, with an elaborate ceremony climaxed with the passage of the first passenger train through the bore.[4]

The Denver and Salt Lake Railroad eventually ended at the town of Craig in northwestern Colorado where it benefited from coal mines in the area, but it never reached Salt Lake City. That was left up to the Denver & Rio Grande Western Railroad, which purchased a majority interest in the Denver and Salt Lake Railroad in 1932. To make the most efficient use of the Moffat Tunnel, the D&RGW decided to end its roundabout journey to Pueblo and Leadville by constructing the Dotsero Cutoff, a 38-mile cutoff from the old D&SL at Bond, Colorado, to Dotsero. This saved 175 miles from Pueblo, but the cutoff was not completed until June 16, 1934. The D&RGW's route from Dotsero then followed the Colorado River west through

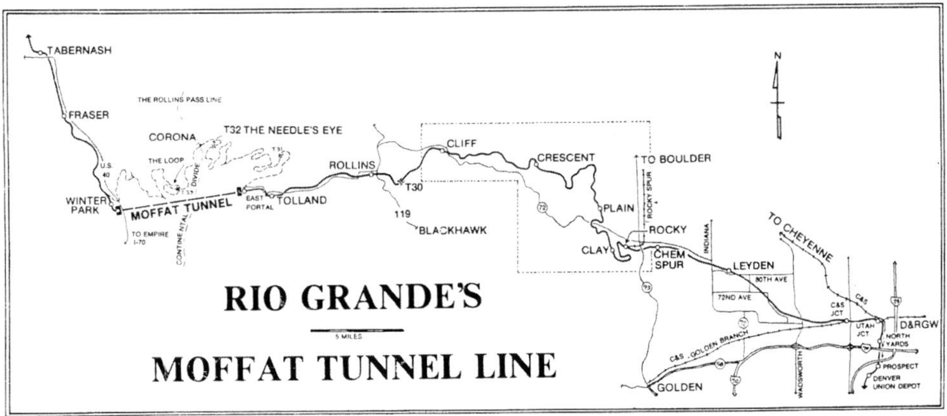

TUNNEL LIST

Denver to Winter Park

No.	Milepost	Location	Length
1	23.43	Coal Creek	299 feet
2	25.44	Plainview	516 feet
3	25.79	Plainview	369 feet
4	26.00	Plainview	174 feet
5	26.11	Plainview	585 feet
6	26.35	Plainview	536 feet
7	26.63	Plainview	208 feet
8	26.91	Plainview	753 feet
9	27.00	Plainview	0 feet
10	27.42	Plainview	1572 feet
11	27.81	Crescent	238 feet
12	27.93	Crescent	429 feet
13	28.14	Crescent	312 feet
14	28.28	Crescent	434 feet
15	28.46	Crescent	427 feet
16	28.72	Crescent	698 feet
Old 17	29.00	Crescent	0 feet
17	29.50	Crescent	1730 feet
18	29.97	Crescent	238 feet
19	32.11	Crescent	1055 feet
20	32.45	Crescent	460 feet
21	32.71	Crescent	667 feet
22	32.98	Crescent	180 feet
23	33.20	Crescent	1622 feet
24	34.10	Crescent	812 feet
25	34.61	Pinecliff	639 feet
26	35.22	Pinecliff	295 feet
27	35.72	Pinecliff	643 feet
28	36.02	Pinecliff	0 feet
29	36.38	Pinecliff	78 feet
30	40.47	Rollinsville	257 feet
Moffat	50.18	East Portal	6.21 miles
Tunnel	56.39	Winter Park	

Tunnels 3-4-6-24 exist today as built. Tunnel 10 had a disastrous fire in 1943 and Tunnel 20 had a fire in 1905. Tunnel 9 caved in when almost completed but due to pressure on timbering was abandoned and bypassed in 1903. Never was operated. Old Tunnel 17 caved in when partially completed and abandoned in 1903. Now replaced by a cut which destroyed the old tunnel. Tunnel 28 was daylighted in 1951 due to sloughing of the sidewalls.

The milepost locations of the thirty tunnels from Denver to Winter Park. Tunnels 31, 32, and 33 were on the original line over Rollins Pass.

The route from Denver to Winter Park. Note the location of tunnels 31, 32, and 33 over Rollins Pass (Corona Station).

Tunnel 32—the Needle's Eye—on the Rollins Pass route is seen here in November 1987. The railroad carved into the mountainside, and the drop-off at the edge was very dramatic.

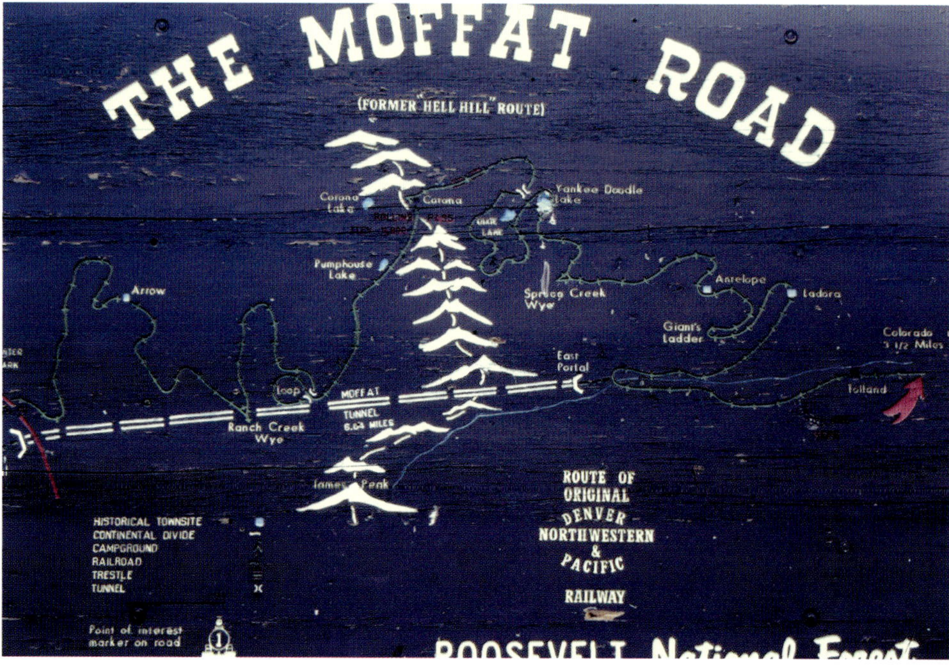

In the 1980s, Roosevelt National Forest posted historic signs and markers along the Rollins Pass road, which could be driven by automobiles before the cave-in of the Needle's Eye tunnel.

The completion of the Moffat Tunnel (bottom center) under James Peak eliminated the treacherous rail route over Rollins Pass. Note the container freight train at the lower left.

Glenwood Canyon, on to the Colorado border, and then northwest to Salt Lake City, completing David Moffat's dream.

Snow and ski trains to Winter Park have always been able to use the Moffat Tunnel Route. Fortunately, they did not have to run over the treacherous Rollins Pass route:

> The first official "ski train" ran … on Sunday, February 9, 1936. The Denver Rocky Mountain News sponsored the train and publicized it highly in the newspaper. Over 2,000 tickets were sold, and the train hauled passengers out to the 25th annual U.S. Western Ski Tournament and Winter Sports Carnival at Hot Sulphur Springs [about 30 miles northwest of Winter Park].[5]

The Winter Park Ski Area was established in 1940 after Denver Improvements and Parks Manager George Cranmer had arranged for the purchase of 89 acres for the ski resort base as a Denver Park:

> Winter Park opened on January 28, 1940, with $1.00 lift tickets to three ski runs served by a T-bar lift. The run at the top of the mountain was named 'Cranmer' in honor of Winter Park's most enthusiastic supporter.[6]

After World War II, the Denver & Rio Grande Western Railroad resumed ski trains to Winter Park in 1948. The trains usually ran from Christmas week and on weekends through March of the following year.

The Silver Sky was positioned at the tail end of the Ski Train for the Rocky Mountain Railroad Club's excursion on Saturday, March 28, 1987. It was a remnant of the famous "Zephyr" streamliner trains which ran from Chicago to California. The name "Zephyr" was chosen by Ralph Budd, supposedly after reading Chaucer's *Canterbury Tales*, which described Zephyrus as "the Greek personification of the west wind, a force of renaissance and change. The Anglo-Saxon derivative would be Zephyr."[7]

The route of the Zephyr was shared by three railroads: the Chicago, Burlington & Quincy Railroad handled the train from Chicago to Denver; the Denver & Rio Grande Western Railroad covered the route from Denver to Salt Lake City; and the Western Pacific Railroad traversed the final leg from Salt Lake City to Oakland. Each railroad was required to purchase a number of the silver streamliner passenger cars according to the total mileage of their portion of the route, and they were required to use their own motive power on their portion. Thus, the California Zephyr train sported silver cars all the way from Chicago to Oakland, but the train had different locomotives on each portion of the route. "The train soon became one of the most popular in the United States."[8]

The California Zephyr ran from March 21, 1949, to March 21, 1970. Heavy financial losses due to competition from airlines and automobiles caused the demise of the Zephyr. On March 22, 1970, the Denver & Rio Grande Western Railroad continued its portion of the Zephyr, while the other original two legs of the journey were abandoned. This train, known as the Rio Grande Zephyr, ran from Denver to Salt Lake City until April 24, 1983. Finally, Amtrak began full California Zephyr service from Chicago to Oakland from July 16, 1983, to the present using its own domeliner cars.

Industrialist Philip Anschutz purchased the D&RGW Railroad in November 1984. Anschutz was a railroad buff, and he was intent on keeping the Ski Train operating to Winter Park. He sold the decrepit old Northern Pacific passenger cars to the Napa Valley Wine Train operation in California, then purchased seventeen Tempo cars from VIA Rail Canada. The VIA cars began arriving in Denver in November 1987. "Fourteen of them were painted in Rio Grande gold and silver with black lettering and striping."[9] The new Ski Train began running on January 2, 1988. "Three Zephyr cars—Silver Bronco, Silver Shop, and Silver Sky remained under Rio Grande ownership for its own specials."[10] When the Rocky Mountain Railroad Club took its excursion on the Silver Sky to Winter Park on March 28, 1987, the car was under Anschutz's ownership, bringing up the rear of the old Northern Pacific cars of the Ski Train.

Right: The club's tail-plate is seen at the end of the Silver Sky parked at Winter Park.

Below: The Silver Sky, formerly part of the California Zephyr, rests at the end of the Ski Train at Winter Park as skiers head to the slopes.

Denver & Rio Grande engine 3095 substitutes for a dead Amtrak engine left behind for repairs in Denver. The Amtrak California Zephyr is coming out of the West Portal of the Moffat Tunnel into Winter Park on March 28, 1987.

Amtrak pulls into Winter Park. The train did not stop here. Its scheduled stop was at Fraser, a few miles to the west.

The Rocky Mountain Railroad Club chartered a bus to allow members to follow the Silver Sky and Ski Train to its turnaround at the Tabernash wye. The train here is northbound, with the view to the west.

The Ski Train has been wyed and is ready for its return trip to Denver on March 28, 1987. Two Rio Grande engines are headed to the West Portal of the Moffat Tunnel.

The Ski Train is loading skiers for the return to Denver, with the Silver Sky bringing up the rear.

2

The Manitou and Pikes Peak Cog Railway: May 9, 1987

The Rocky Mountain Railroad Club was formed on March 30, 1938:

> Carl Hewett called the first meeting of the Rocky Mountain Railroad Club for an intimate gathering of about twenty in the basement of the Union Pacific freight house at 19th and Wynkoop Streets in downtown Denver … He could not have known what a far-reaching influence the club would have.[1]

The club members "agreed to meet periodically to share stories, take trips, and show photographs and films of Rocky Mountain region railroad activities."[2] The very first excursion of the club was on the Manitou and Pikes Peak Cog Railway on August 27, 1939.

The charter trip for the club on the Cog Railway was arranged by Carl Hewett with a fare of $3.50, and about twenty members and guests rode on the train, powered by Manitou and Pikes Peak steam engine no. 3. Engine no. 3 was built by the Baldwin Locomotive Works in May 1890:

> In [April] 1890 the first rack-rail locomotive on the Abt system was constructed for the Pike's Peak Railroad, and during this year and 1893 four locomotives were built for working the grades of that line, which vary from 8 to 25 percent. One of these locomotives, weighing in working order 52,680 pounds, pushes 25,000 pounds up the maximum grades of one in four.[3]

The boiler on engine no. 3 (as in all of the first four engines built by Baldwin) was offset sixteen degrees to keep it level on the steep inclined grades of the railroad. Most locomotives pull rail cars behind them, but these Baldwin steam engines pushed the Cog Railway passenger car up the mountain and helped control its descent down the mountain. The rack rail system on a cog railway consists of a toothed rail between the running rails. The locomotive is fitted with a cog wheel (between its running wheels) that meshes with the rack rail. This allows trains to operate on steep grades above 10 percent. The average grade on the Pikes Peak Cog Railway is 12 percent but there are grades up to 25 percent. The Abt rack rail system was developed by Carl Abt, a Swiss locomotive engineer, in 1882.

The Pikes Peak Cog Railway was built by Zalmon Simmons. Simmons was surveying a route up Pikes Peak in 1888 for a telegraph line to the top. He rode a mule to the top and after a rather miserable trip decided that a railroad to the top would be a much better way to get there. Construction of the cog railway started in 1889 and was completed in 1891. The line is 8.9 miles long, and the first train reached the summit of Pikes Peak on June 30, 1891. In 1926, Spencer Penrose, owner of the Broadmoor Hotel in Colorado Springs, purchased the Cog Railway from Simmons. In 2011, industrialist Philip Anschutz purchased the Broadmoor and the railway. Thus, the name of the railroad today is the Broadmoor Manitou and Pikes Peak Cog Railway. The railway does not start at the Broadmoor however; it starts at the depot in Manitou Springs, elevation 6,320 feet, and finishes at the top of Pikes Peak, elevation 14,115 feet.

Replacement of the steam locomotives of the Cog Railway began in 1939. The Rocky Mountain Railroad Club was fortunate to have their chartered car pushed up the mountain in 1939 by a steam locomotive before they were phased out. Several diesel electric locomotives were then purchased from General Electric for the haul up Pikes Peak. The railroad started switching over to a fleet of self-propelled diesel railcars in 1964, purchased from Swiss Locomotive and Machine Works. These early units could carry seventy-eight passengers. In the 1970s, tourist rides on the Cog Railway dramatically increased. Passing sidings were built at Minnehaha and Windy Point, allowing up to eight trains to travel up the mountain per day. In 1983, Swiss Locomotive and Machine Works delivered a new two-car unit that could carry 214 passengers. It was this set that the author rode on his May 9, 1987, excursion with the Rocky Mountain Railroad Club.

By the 2000s, the rails and equipment of the Cog Railway were showing considerable wear and tear. When the railroad closed for maintenance on October 29, 2017, it was discovered that serious repairs were needed. By March 2018, the Anschutz Corporation decided to close the railroad indefinitely, completely tear up all the tracks and replace them, rebuild older rail cars, and purchase three new train sets from Stadler Rail of Switzerland. Each of these new train sets consisted of a diesel-electric locomotive, two coaches, and a control car. The completely rebuilt Cog Railway reopened on May 20, 2021. The new "Cog Railway is serviced by three new trains that can carry up to 263 passengers per trip and four refurbished railcars that carry 214 passengers each."[4]

With up to ten round-trips per day in the summer months, the number of passengers on top of Pikes Peak easily overwhelmed the old visitor center there. The city of Colorado Springs controls the summit and has built a completely new visitor center which can handle the crowds. Fortunately, the tradition of baking donuts at 14,115 feet has continued in the new visitor center restaurant, and passengers can delight in this refreshment today, just as passengers of the Rocky Mountain Railroad Club did in 1987.

The Abt rack rail system is illustrated with the running wheels of the locomotive on the outside and the cog wheel in the middle which meshes with the rack rail. (*Photograph from Wikipedia Creative Commons*)

This William Henry Jackson photograph (*c.* 1900) shows Baldwin steam locomotive no. 3 with its offset boiler ready to push the rail car up Pikes Peak. (*Library of Congress, LC-D4-13812*)

Manitou Springs Depot of the Cog Railway on October 21, 2012. (*Photograph by Milan Suvajac, Wikipedia Creative Commons*)

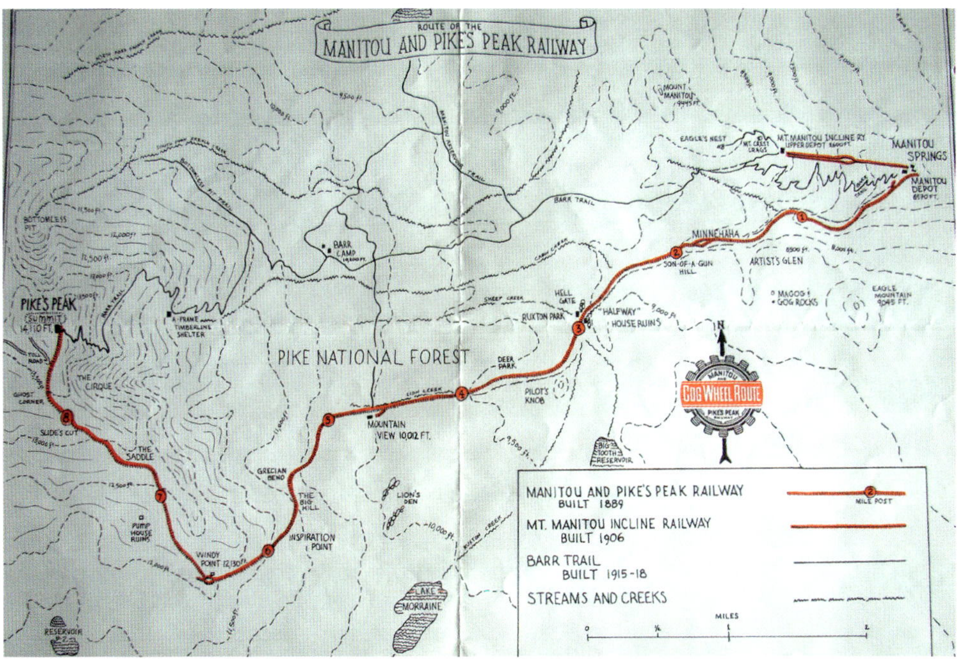

The 8.9-mile route of the Cog Railway from Manitou Depot to the summit of Peaks Peak. Note the passing sidings at Minnehaha and Windy Point. (*Rocky Mountain Railroad Club trip brochure*).

Left: Passing sidings allowed the Cog Railway to operate more trains daily. This is the Minnehaha Siding.

Below: Snow encroaches at the Windy Point Siding on the club's excursion May 9, 1987.

Above left: Engine no. 16 belches diesel smoke as it heads uphill through "Slide's Cut" about 1/2 mile below the summit of Pikes Peak.

Above right: This runby at Slide's Cut was quite a treat for members of the club—membership has its advantages.

Below left: Passengers get ready to depart the train at Pikes Peak summit.

Below right: Club members climb down into the snow on top of Pikes Peak. The Abt rack rail is clearly seen at bottom right.

Above left: Engine no. 16 parked at the summit of Pikes Peak. Intrepid travelers could tromp through the snow to the viewing telescopes at left for an incredible view of Colorado Springs below.

Above right: The yellow and black "Bumblebee" work engine was parked at the summit on the siding next to engine no. 16.

Jason Butler, oblivious to cold and snow, enjoys the view from the top on May 9, 1987.

3

The Cadillac and Lake City Railway: September 19, 1987

What makes Rocky Mountain Railroad Club excursions unique is the vast diversity of experiences offered to members. From the 2-foot-gauge Cripple Creek and Victor Railroad, to the 3-foot narrow gauge of the Durango and Silverton Railroad, to the standard gauge of the Union Pacific across southern Wyoming, to hikes along Tennessee Pass and Boreas Pass, the club has offered a variety of trips sure to interest its members.

Perhaps one of the stranger club trips was on the Cadillac and Lake City Railway, named after two cities in Michigan. One wondered what this had to do with Rocky Mountain rail history. It turns out that the Cadillac and Lake City Railway had branches in three separate states: Michigan, Kansas, and Colorado. The railroad began operations in 1964 over former Pennsylvania Railroad tracks between Cadillac, Michigan, and Lake City, Michigan—a total of 21 miles. In July 1981, the Cadillac and Lake City Railway began operations over former Chicago Rock Island & Pacific Railroad tracks between Limon, Colorado, and Colby, Kansas—a total of 171 miles. This service was turned over to the Kyle Railroad in May 1984. The Cadillac and Lake City Railway then began freight service (and occasional passenger excursions) on the former Rock Island tracks from Falcon, Colorado to Limon—a total of 61 miles. This is the segment traversed by the Rocky Mountain Railroad Club on September 19, 1987.

Thus the story of the Cadillac & Lake City in Colorado is steeped in the history of the Chicago Rock Island & Pacific Railroad, which originally built the tracks in 1888, a total of 177 miles from the Kansas–Colorado border to Colorado Springs:

From the Colorado state line the track extended west across the plains to the new town of Limon where it crossed the Kansas Division of the U.P. (Union Pacific). A short distance farther it swung toward the southwest and continued to Roswell, just north of Colorado Springs. There, on October 14 the tracklayers joined the rails with the main line of the D&RG (Denver & Rio Grande) ... The first

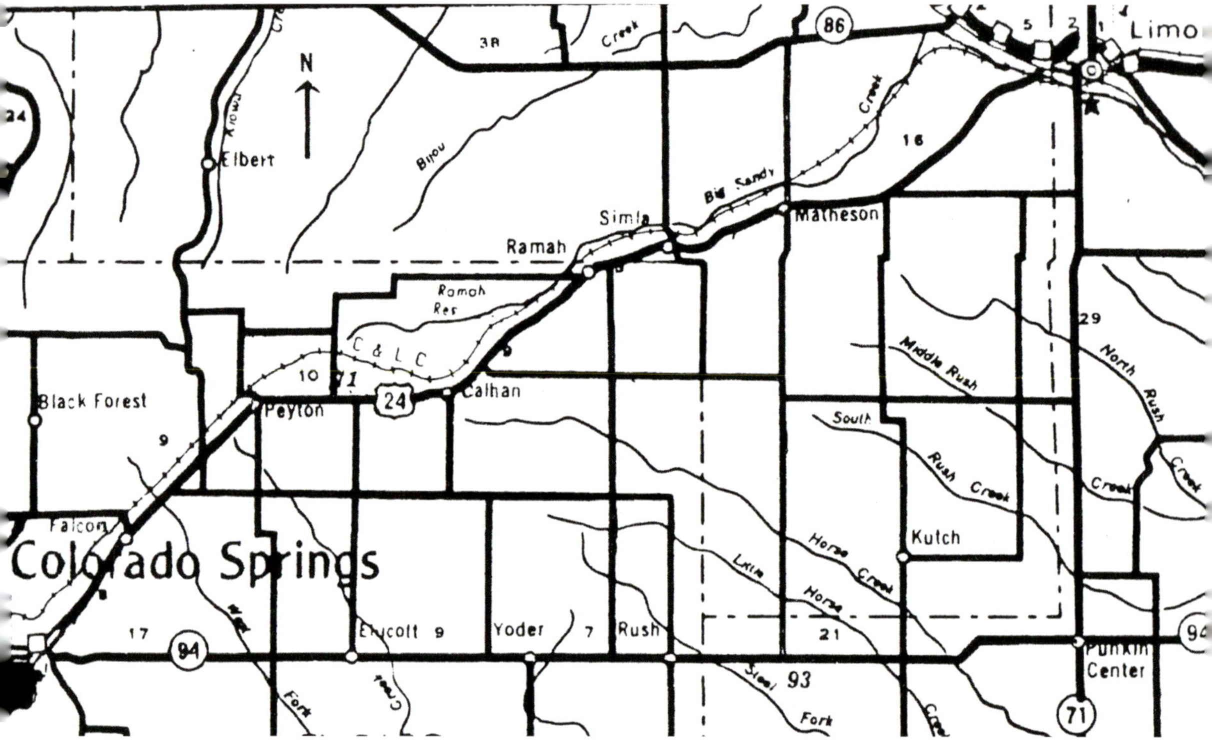

The 61-mile route of the Cadillac & Lake City Railway from Falcon to Limon is shown on this map. There was a depot at Calhan and Limon, and a bridge over Big Sandy Creek north of Matheson. (*Rocky Mountain Railroad Club map*)

Pikes Peak looms in the middle background as the Cadillac & Lake City excursion of the Rocky Mountain Railroad Club heads east from Falcon in September 1987.

Right: The club's tail-plate adorns the rear of the train on the Mountain View car.

Below: The Cadillac & Lake City Railway excursion passes by Tiptop, the highest point on the Chicago Rock Island & Pacific Railway.

Rock Island passenger train arrived on November 4 and departed the following morning on an established schedule for mid-western points ... The following year an agreement was made with the U.P. for the use of its tracks between Limon and Denver.[1]

Limon was apparently named for the Rock Island foreman in charge of building the tracks:

The town was established as a camp for the Rock Island Railroad; It was known as Limon's Camp for the foreman, and later as Limon's Junction for the meeting of the Rock Island and Union Pacific railroads. The spelling "Lyman Junction" is also recorded.[2]

In the 1930s, the Chicago Rock Island & Pacific Railroad became interested in boosting its passenger traffic after suffering declining freight revenues during the Great Depression. As America came out of the Depression, people wanted to travel, particularly to Colorado with its towering mountains and cooler summer temperatures.

Recognizing a possible passenger bonanza on its line to Colorado Springs (and connections to Denver), the Rock Island in 1937 purchased the following:

6 passenger diesels ... [and] 20 stainless steel streamlined passenger cars of various classes. The name 'Rocket' was chosen for these new trains. The Rock Island Rockets. This sounded good to the ear, and putting it into type looked good to the eyes.[3]

The Rock Island immediately began placing its Rockets into passenger service in the mid-west with a Rocket from Chicago to Peoria. In 1939, the Rock Island decided to "go ahead with a new streamlined Rocky Mountain Rocket":

[This] would offer to the public the only main-line hot-shot between Chicago and Colorado Springs ... [it] would split the Rocket at Limon and send half of it to Denver and the other half to the foot of Pikes Peak. Vacationists heading for the Pikes Peak area and the dude ranches beyond would no longer have to change trains at Denver to get to their destination.[4]

The Rocky Mountain Rocket ran until 1966 when it was discontinued, ending Rock Island passenger service in Colorado. Freight operations continued in Colorado until 1980, when the Chicago Rock Island & Pacific Railroad ceased all its operations.

When the Rocky Mountain Railroad Club took its excursion on the Cadillac and Lake City Railway from Falcon to Limon in September 1987, the train operated at a much more leisurely pace than the streamlined Rocky Mountain Rocket. Three to four hours were allotted to the one-way 61-mile journey, but that would allow for three photo runbys on the eastbound leg, and two runbys on the westbound return trip. It would also allow for picking up and distributing some freight cars along the

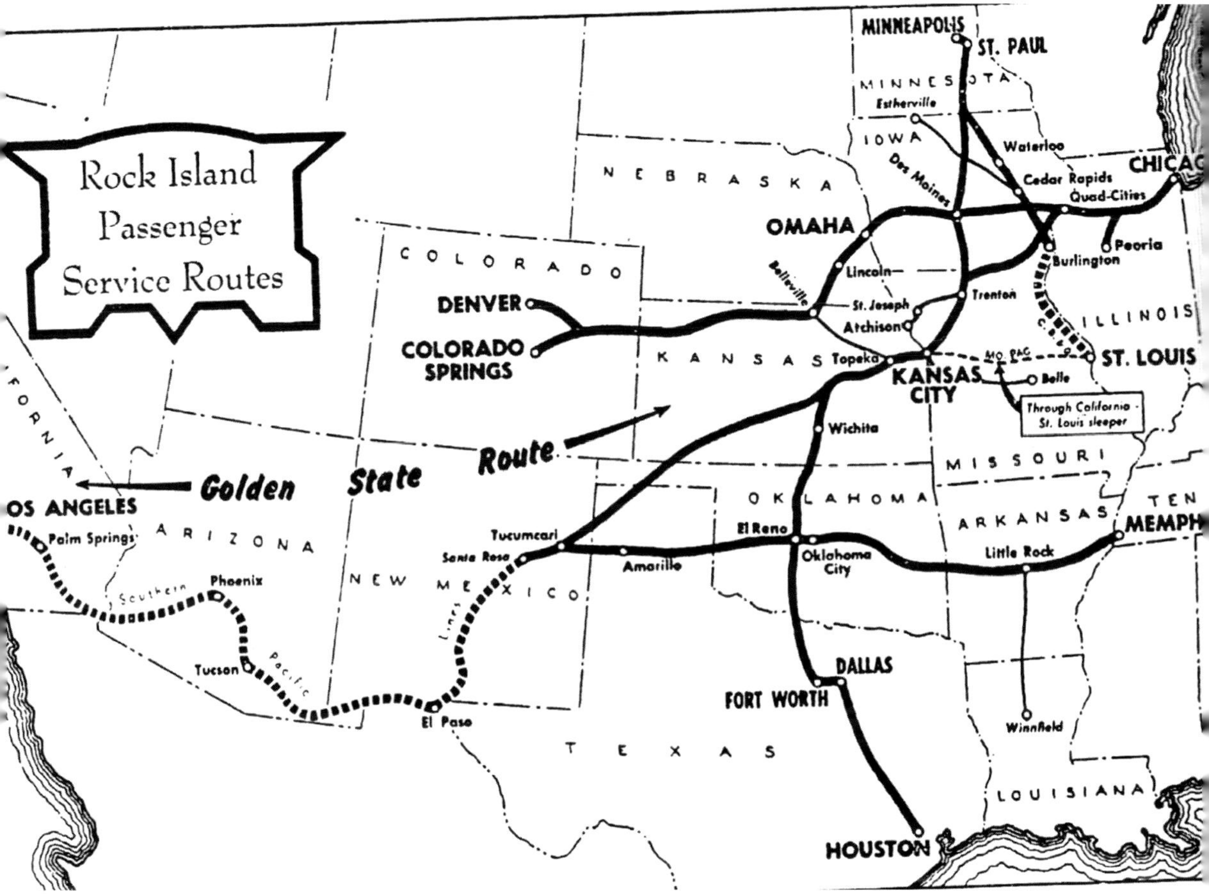

Map of the Rock Island passenger routes shows the route between Chicago, Omaha, and Colorado Springs/Denver. (*Author's collection*)

This Chicago Rock Island & Pacific passenger train steams east, leaving Pikes Peak and Colorado Springs behind, *c.* 1910. (*Library of Congress, LC-D4-43492*)

way. The equipment was a very mixed bag of used cars from various railroads. The club brochure for the trip gave some historical background to each car as noted below.

Locomotive power for the excursion was provided by a CF7 engine originally owned by the Atchison Topeka and Santa Fe Railway. It was sold to the Cadillac & Lake City in the mid-1980s. The first car behind the engine was the "Smokey Hill River." This car was built as a coach in 1912 for the Chicago Burlington and Quincy Railroad:

> They rebuilt it into a combine car in 1961-1962 with one baggage section, electrical generator section, and a coach seating section with 45 spaces. The car provided head-end power on the Q's Chicago commuter trains.[5]

The Cadillac & Lake City acquired the Smokey Hill River in 1985.

The second car behind the locomotive was the cafe car. It was built as a sixteen-section drawing room sleeper in 1914 for the Pullman Company, named the Curena. In 1957, it was sold to the Great Northern Railway and converted to a lunch counter car. It was sold to the Cadillac & Lake City in 1984, which continued to use it as a cafe car.

A Rock Island caboose and a Santa Fe caboose sandwich an old Amtrak car on this siding of the Cadillac & Lake City Railway just east of Falcon.

The mixture of old railroad passenger cars used on the Cadillac & Lake City Railway excursion is noted by the different paint schemes.

The Smokey Hill River car is directly behind the locomotive. On the nose of the locomotive are the letters "CLC" for the Cadillac & Lake City Railway.

The cafe car at the right was numbered CLK-101 for Cadillac & Lake City. It was the second car behind the locomotive, as club members stretched their legs at the Calhan depot.

The third car behind the locomotive was the "Silver Pine." This was a streamlined car built in late 1948 for the California Zephyr at a cost of $148,810. It was originally a sixteen-section sleeper with upper and lower berths for thirty-two passengers. In the mid-1960s, it was rebuilt into a forty-eight-seat coach. On the California Zephyr, it was usually placed before the first dome car. Amtrak purchased the car in 1983, and it eventually made its way to the Cadillac & Lake City.

The fourth car behind the locomotive was an Amtrak car for which the club provided no information. Finally, the fifth car behind the locomotive was the "Mountain View." This car was built in June 1949 by the Budd Company as a five double-bedroom/buffet lounge car for the New York Central Railroad, used in service between New York and St. Louis. It had several different owners over the years, and was eventually modified to business car use with the observation area at the rear. The Cadillac & Lake City acquired the car in 1985.

While the train was certainly a mixed bag of cars, the club's excursion was quite enjoyable. It turned out to be a delightful fall day, with views of Pikes Peak to the west and the undulating Colorado prairie to the east. The return trip from Limon in the late afternoon was blessed with a glowing sunset behind Pikes Peak, which made for great photographs and a memorable ending to the journey. Unfortunately, the Cadillac and Lake City Railway ceased operations in Colorado in 1989. The Rock Island tracks were removed in the early 1990s. Remaining remnants of the line were the depot at Calhan and the bridge over Big Sandy Creek near Matheson.

The Silver Pine was a streamlined car from the California Zephyr, and it rode in the third position behind the locomotive.

The Silver Pine rests on a trestle on its eastbound journey to Limon.

Club members return to the Mountain View car after a photo runby.

The Mountain View brings up the rear of the Cadillac & Lake City Railway excursion to Limon, offering passengers a relaxing view from the curved rear observation windows.

Having picked up a couple of freight cars, the club excursion continues east over the bridges at Big Sandy Creek, north of Matheson.

Passenger cars clear the Big Sandy Creek bridge heading east to Limon.

The two freight cars are shown ahead of the Smokey Hill River car as the excursion train continues eastward to Limon.

Club members depart the train at the Limon depot, heading for a lunch of BBQ brisket sandwich, potato salad, coleslaw, baked beans, and cake.

The train heads into the yard at Limon for more switching of freight cars during the lunch break.

Heading west after lunch, there are two freight cars behind the locomotive and then the Mountain View car as the passenger cars were not turned in Limon.

The train continues west, having shed one of the freight cars. The passengers in the Mountain View observation car now have an excellent view of the boxcar directly ahead of them.

4

Back-to-Back Narrow-Gauge Excursions: June 23–25, 1988

The Rocky Mountain Railroad Club's fiftieth anniversary celebration in 1988 was a big deal. A massive number of rail excursions were planned for the last two weeks of June, including trips in Wyoming, Colorado, and New Mexico, with planning beginning several years in advance. The club's anniversary brochure stated:

> Our plans include articulated, mainline steam, historic and colorful narrow gauge operations, cog railroading to the top of America's most famous mountain, a shortline trip across the high plains of Colorado, a visit to one of the most historic mining districts in the United states and field trips along the grades of the famous narrow gauge 'ghost railroads' of the silvery mountains of southwestern Colorado. The itinerary has been formulated to allow participation in all or part of the festivities. The scheduling of these activities will provide an outstanding opportunity for out-of-state members and friends to experience an exceptional variety of rail experiences during a single visit to the region!!! Not to be forgotten are the many other attractions that Colorado has to offer plus the endless miles of scenic beauty that await the traveler.[1]

So if a person had two weeks of vacation time and at least $251.50 (cost of the Union Pacific excursion not yet included), one could participate in all of the festivities.

The author and his ten-year-old son Jason were fortunate to participate in the "Back to Back Narrow Gauge Excursions" from June 23 to June 25 on the Cumbres & Toltec Scenic Railroad (June 23) and the Durango & Silverton Narrow Gauge Railroad (June 25). A day was left open on June 24 to travel by automobile from Chama, New Mexico, to Durango, Colorado, along the old Denver & Rio Grande Railroad right-of-way. Recollections of that trip resulted in the author's 2022 book *Tracking the Narrow Gauge from Chama to Durango* (see bibliography).

Several excellent books have been written about both the Cumbres & Toltec and the Durango & Silverton railroads, so only a brief mention of the history of each will be

ROCKY MOUNTAIN RAILROAD CLUB
50TH ANNIVERSARY CELEBRATION - - JUNE 18 - JULY 2, 1988
Ticket order form

Notes	Event No.	Day	Date	Event	No. of Tickets x	Price Each	=	Total Amount
(A)	1.	SAT.	6/18	Union Pacific No. 3985	____ x	$ XXX.XX	=	$____
	2.	SUN.	6/19	Georgetown Loop	____ x	$ 25.00	=	$____
	3.	MON.	6/20	Manitou & Pikes Peak	____ x	$ 18.00	=	$____
	4.	MON.	6/20	Flying W Ranch Dinner	____ x	$ 9.50	=	$____
	5A.	TUES.	6/21	Cadillac & Lake City	____ x	$ 30.00	=	$____
	5B.	TUES.	6/21	Cripple Creek Tour	____ x	$ 8.00	=	$____
(B)	6	TUES.	6/21	Guiseppe's Restaurant	____ x	0.00	=	$____
	7	THUR.	6/23	Cumbres & Toltec Scenic	____ x	$ 60.00	=	$____
(C)	8	FRI.	6/24	Chama/Durango Field Trip	____ x	0.00	=	$____
	9	FRI.	6/24	Evening Banquet	____ x	$ 16.00	=	$____
	11	SAT.	6/25	Durango & Silverton NG RR	____ x	$ 60.00	=	$____
(D)	12	SUN.	6/26					
	thru	SAT.	7/2	San Juan Sojourn Field Trip	____ x	$ 25.00	=	$____

TOTAL TICKET ORDER = $

(E) 50% DEPOSIT WITH TICKET ORDER (E) = $

BALANCE DUE BY APRIL 1, 1988 = $

Notes: (A) Price to be determined. (Please refer to trip discription in this flyer.)
    Check this box if you wish to have a 3985 trip flyer mailed to you when it is available. --- >
(B) Reservations only. Payment to be handled individually following dinner.
(C) Reservations only, no cost involved.
(D) $25.00 per adult, no charge for children. Fare includes informative handout for the week's intinerary plus the services of a guide.
(E) Include check or money order made payable to the Rocky Mountain Railroad Club and send your order to the Club's address at P.O. Box 2391, Denver, Colorado 80201. PLEASE DO NOT SEND CASH.

Above: Rocky Mountain Railroad Club order form for tickets for the 1988 fiftieth anniversary excursions.

Right: Club tail-plate for the Cumbres & Toltec Scenic Railroad excursion on June 23, 1988.

Club members gather at the Chama station, wondering where the locomotives are for the club's excursion train.

given here. For more information, readers are referred to Doris Osterwald's excellent two books on those railroads (as noted in the bibliography).

As the Denver & Rio Grande Railroad (D&RG) was building a narrow-gauge route west across southern Colorado in 1880, its goal was to reach Durango and then the mines at Silverton. The railroad reached Antonito, Colorado, on March 30, 1880—a distance of 28.9 miles south of Alamosa. From Antonito, the D&RG turned west to head over Cumbres Pass to Chama, New Mexico, reaching that sheep camp in December, 1880. From Chama, the D&RG continued west across the Jicarilla Apache Reservation in northern New Mexico, then crossed back into Colorado, covering 111 miles from Chama to Durango. The railroad had started construction in its new town of Durango in 1880, and tracks reached the town on July 27, 1881. It took another year for the D&RG to complete the 45 miles from Durango north to Silverton, reaching Silverton in July 1882. The D&RG continued service to Durango until December 9, 1968, "when the recently overhauled [engine] #481 hauling rolling stock needed for the Silverton Branch made the final trip from Alamosa to Durango."[2]

> The D & RG continued to operate the Silverton Branch as an isolated segment of its once vast narrow gauge system until 1981, when the line was purchased by Charles E. Bradshaw, Jr., and renamed the Durango & Silverton Narrow Gauge Railroad.[3]

While the Durango–Silverton tourist train continued operation, efforts were turned to the preservation of the abandoned line from Antonito to Chama for a tourist train.

Colorado and New Mexico were both interested in preserving this line, and each state formed a railroad authority to accomplish the task:

> The New Mexico and Colorado Railroad Authorities purchased the 64 miles of track between Chama and Antonito for $547,120 in July 1970 [from the D&RG]. On September 1, 1970 the first of three large shipments of engines, rolling stock, and non-revenue equipment was delivered at Antonito to the Cumbres & Toltec Scenic Railroad.[4]

The Cumbres & Toltec Scenic Railroad has continued its tourist train operation in the summer months to the present day, thanks to the efforts of hundreds of volunteers. Thus, two segments of the former Denver & Rio Grande Railroad are preserved today, with a 111-mile gap of abandoned line between Chama and Durango. The author rode the Rocky Mountain Railroad Club's "Back to Back Narrow Gauge Excursions" on the Cumbres & Toltec Scenic Railroad on June 23, 1988, and the Durango & Silverton Narrow Gauge Railroad on June 25, 1988, with the 111-mile gap covered by automobile on June 24.

The club's excursion on the Cumbres & Toltec was scheduled to depart at 7:30 a.m. on June 23 with double-headed steam locomotives. Club members were instructed to arrive at the Chama station at 7 a.m., but upon arrival, there were no locomotives in sight to attach to the club's charter train. The train consist was a special lash-up of ten freight cars with a caboose followed by ten passenger cars. This would allow photographs of just the two locomotives and freight cars, resembling an old-time D&RG freight run. Information trickled down that one of the locomotives for the trip had broken down, so a loco had to be deadheaded to Chama from Antonito. The second locomotive had been involved in motion picture filming on the line to the east and was late arriving back in Chama. Finally, at 8:30 a.m., the two locomotives were in place and headed east with the club's excursion train. At Cumbres Pass, the lead locomotive was uncoupled and deadheaded to Antonito. A lunch stop was arranged at Osier. A new $600,000 dining facility was opened at Osier on June 30, 1989, just one year after the club's visit.

After lunch, the train proceeded on to the Big Horn Wye and Siding. The siding at Big Horn is 1,184 feet long and will hold twenty-eight cars. Once turned on the wye, the locomotive could then pick up the cars from the siding for the trip back to Chama. Several photo runbys were arranged throughout the trip, and it made for a spectacular day. The train arrived back in Chama at 8:30 p.m. in a driving rainstorm, producing several inches of standing water in the RV park next to the train track.

On June 24, the club's trip continued with the drive along the D&RG grade from Chama to Durango. There were paved roads from Chama to Dulce, New Mexico, headquarters of the Jicarilla Apache Nation. After that, the roads were dirt and gravel, with a particularly rough section just west of Dulce beginning at Navajo. At Navajo, the D&RG bridge over the Navajo River is preserved, as is the water tank. Crossing into Colorado, the ghost town of Juanita was observed, and just west of Juanita, the spectacular three-span D&RG bridge over the Navajo River still exists. Next

Finally departing Chama at 8:30 a.m., the Cumbres & Toltec's double-headed locomotives steam east out of Chama.

The Lobato trestle would not tolerate the weight of two locomotives, so the lead engine had to be uncoupled and driven across separately.

Club members search for the approach of the train for the first photo runby, just beyond the Lobato trestle which can be seen in the mid-distance.

Cumbres & Toltec engines number 488 and 489 steam up the grade east toward Cumbres Pass with freight cars behind in this special consist for the club.

The double-header climbs toward the spectacular Windy Point, just below Cumbres Pass.

The train has pulled into Cumbres Pass, elevation 10,015 feet, with the D&RG gold-colored section house on the right.

Above left: Continuing east toward Cascade Trestle, the train has lost one engine that was uncoupled at Cumbres Pass and returned to duty elsewhere on the line.

Above right: The spectacular Cascade Trestle draws "oohs" and "aahs" from the rail fans.

Engine no. 489 steams toward Big Horn Wye, with its trailing freight and passenger consist.

Club members hop off the train at Big Horn for a photo runby.

The train works its way through the Big Horn Wye in preparation for the return trip west to Chama.

The full extent of the club's special train with its ten freight cars, caboose, and ten passenger cars is shown on Tanglefoot Curve, just east of Cumbres Pass.

stop was the ghost town of Pagosa Junction, which had one family still living there on land leased from the Ute Indian Reservation. Prominent features of the D&RG remained with the section house, pump house, water tank, tracks with a gondola car on a siding, and a bridge over Gato Creek. The Gomez general store was still there in good condition.

> In 2000, Lilliosa Padilla and her son Ray were the only two residents left in Pagosa Junction. Felix Gomez, Lilliosa's father, had closed the doors on the Gomez Store in 1971, leavings its contents intact. Since then, Mrs. Padilla has conducted guided tours of the historic building and its contents.[5]

The Railroad Club was fortunate to have one of these guided tours of the Gomez Store during its trip. Sadly, the Gomez Store is no longer in Pagosa Junction. When the Padilla lease was not renewed by the Ute Tribe in 2000, the store was moved to a museum in Pagosa Springs, where it still sat in 2021.

Next up on the "Back to Back Narrow Gauge Excursions" was the club trip on the Durango & Silverton Narrow Gauge Railroad on June 25. Again, this was to be a double-headed locomotive excursion. Before the train pulled out of the Durango station on time at 9:45 a.m., club members were allowed to explore the roundhouse, car shop, and yard. The roundhouse was the original ten-stall building built in 1881. On February 10, 1989, fire destroyed this roundhouse, so the club was fortunate to observe the original roundhouse in 1988. Within a year of the fire, a new fifteen-stall roundhouse was constructed.

The narrow-gauge train ride to Silverton is one of the most spectacular in the nation as it threads its way into the heart of the San Juan Mountains on the narrow rock ledge above the Animas River at Rockwood. At this point on the club trip, the lead locomotive had to be uncoupled and driven across the ledge. Then the remaining locomotive and the nine passenger cars could proceed across the ledge. Weight limits on the ledge will not permit both locomotives to pass at the same time. On the north side of the ledge, the two locomotives were coupled together again, and the train proceeded to Silverton. Due to the narrow confines of the Animas River Canyon, photo runbys of the train were limited. The club did not want to lose a member slipping into the river. However, there were a couple of nice runbys accomplished at wider spots.

The train arrived on time at Silverton, and club members departed for a ninety-minute lunch break in town. One of the dual locomotives was uncoupled during this time and deadheaded back to Durango. A two-locomotive train was not allowed on the journey back to Durango because the downhill grade and weight of the engines cause excessive speed. Unfortunately, most of the club's journey back to Durango was marred by heavy rain, though the storm did let up about 10 miles north of Durango, allowing passengers an enjoyable open-air view of the wide portion of the Animas Valley. The club's "Back To Back Narrow Gauge Excursions" were spectacular and accomplished with no mishaps to passengers or rail equipment.

On June 24, 1988, club members explored the old Denver & Rio Grande Railroad grade from Chama to Durango. The water tank and bridge over the Navajo River are seen here at Navajo, New Mexico, west of Dulce.

The three-span D&RG bridge over the Navajo River, west of Juanita, Colorado, was a spectacular sight in 1988, and fortunately, it is still there today.

Club members explore Pagosa Junction with its D&RG water tank, rails, siding, and bridge over Gato Creek. The Gomez store is the red building off to the right.

A tour of the interior of the Gomez store, closed for several years prior to 1988, was enjoyed by club members. The store was moved to a museum in Pagosa Springs in 2000.

The club's excursion on the Durango & Silverton Narrow Gauge Railroad on June 25, 1988, again featured double-headed locomotives. Here, engines nos. 478 and 476 cross under the U.S. Highway 550 bridge north of Durango.

Engine 478 has been uncoupled and run across Rockwood Ledge prior to engine 476 leading the nine passenger cars across the ledge.

Passengers gasp at the incredible view down into the Animas River Gorge as the train creeps across Rockwood Ledge.

The two engines have been coupled together again to continue the journey north to Silverton as club members depart the train for a photo runby.

Dual plumes of smoke mark the passage of the engines as the train has descended to river level heading up the Animas Canyon.

A rain shower began at this photo runby as the train was nearing Silverton.

Silverton, with its spectacular mountain scenery, unfolds as the train curves toward town.

The club train halts at the Silverton wye, letting the southbound train led by engine number 480 pass by on its way back to Durango.

Above: The club's double-header steams into Silverton with the avalanche paths clearly seen on the precipitous mountain slope on the right.

Below: Jason Butler poses by the club tail-plate at the rear of the train in Silverton.

5

Amtrak to Glenwood Springs:
April 29–30, 1989

The Rocky Mountain Railroad Club's trip on Amtrak to Glenwood Springs in 1989 was a trip through Colorado rail history—from Denver's Union Station west to the Moffat Tunnel (as described in Chapter 1), to Winter Park, Granby, Hot Sulphur Springs, Byers Canyon, Gore Canyon, Red Canyon, Dotsero, and Glenwood Canyon, Amtrak's California Zephyr winds along a route pioneered by the Denver & Rio Grande Railroad.

The D&RG's narrow-gauge tracks reached Glenwood Springs on October 5, 1887, though it was a longer journey prior to construction of the Moffat Tunnel. Trains had to proceed south from Denver to Colorado Springs and Pueblo, then head northwest to Leadville along the Arkansas River, and on north to Red Cliff.

> Beginning at Red Cliff, to which point the Denver and Rio Grande had been built in the fall of 1881, construction crews began to grade and lay narrow-gauge tracks toward Glenwood Springs … in the summer of 1887.[1]

Financial difficulties for the D&RG prevented construction from 1881 to 1887, but the railroad was spurred on in 1887 by competition from the Colorado Midland Railroad, also hoping to reach Glenwood Springs. From Red Cliff, the D&RG's route was as follows:

> [The] rails followed the Eagle River to its confluence with Grand River [now Colorado River], then down the Grand through Glenwood Canyon to Glenwood Springs. The laying of steel was completed to Glenwood Springs on October 5, a special train arrived that evening, and scheduled trains commenced operating the following day.[2]

The Colorado Midland Railroad did not reach Glenwood Springs until a couple of months later in December 1887.

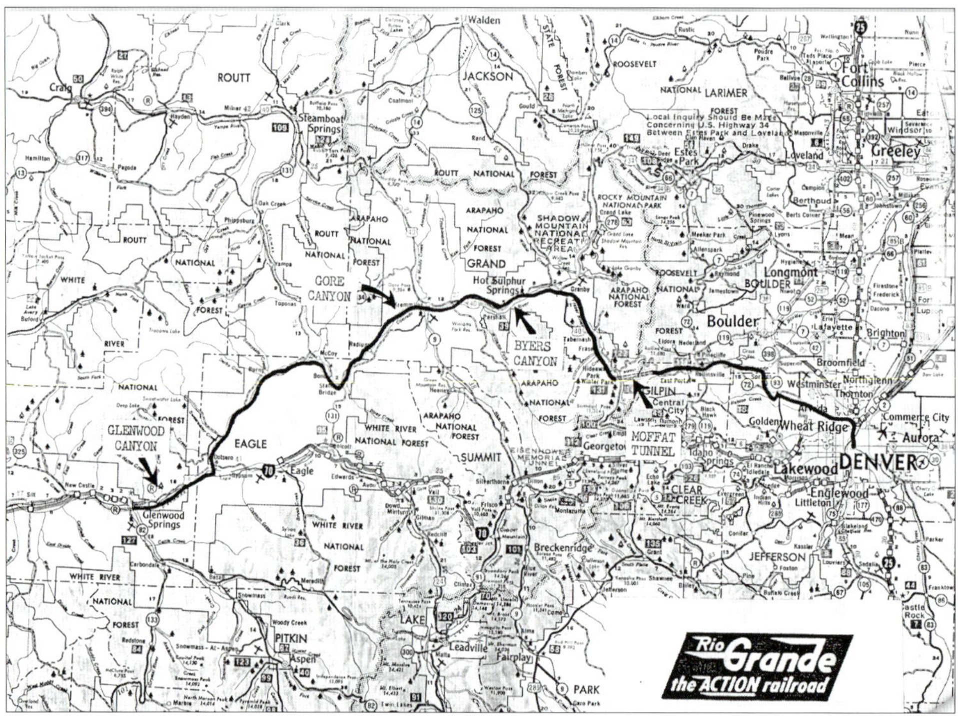

Above: The canyons of Amtrak's California Zephyr route across Colorado are shown in this map provided to members of the Rocky Mountain Railroad Club.

Below: Club members and other passengers gather around the message board in Denver's Union Station to see if the Zephyr will be departing on time.

Entry point for passengers on the California Zephyr at Denver Union Station.

Club members climb aboard Amtrak's California Zephyr at the platform in Denver Union Station.

While the hot springs in Glenwood would later prove to be a great tourist attraction, the railroads were initially interested in the iron ore and coal deposits in the area. Today, Amtrak takes passengers on a leisurely 5.5-hour journey over 185 miles through the Rockies from Denver to Glenwood Springs. It is a great ride in the comfort of the dome-lounge car with spectacular scenery unfolding constantly before the viewer's eyes.

The Railroad Club's trip on Amtrak to Glenwood was a grand affair with approximately forty club members. The author recorded the trip in his journal, and selected passages are reproduced here:

A winter's chill still permeated the air on Saturday morning April 29, as members of the Rocky Mountain Railroad Club filed into Denver's Union Station to board Amtrak's westbound California Zephyr for Glenwood Springs. That cold late April morning, the Zephyr ascended the foothills west of Denver amid lightly falling snow. Crossing the bridge over Coal Creek Canyon and on into Tunnel #1, Amtrak passed through scenes much more reminiscent of winter than spring. On to Pine Cliff and Rollinsville, Tolland and East Portal, bursting through the Moffat Tunnel and out at Winter Park where two inches of fresh snow covered the ground.

The sun began to make brief appearances as the Zephyr entered Fraser Canyon. A few passengers boarded in Granby, and then the Zephyr began flying through canyons named after early Colorado pioneers. Byers Canyon begins just west of Hot Sulphur Springs and is named after William Byers, owner of Denver's first established newspaper, the Rocky Mountain News. Byers also owned some ranch land near Hot Sulphur Springs. The reddish sandstone of Byers Canyon gives way later to the steep gray granite of the spectacular Gore Canyon. While cliffs rise 1,500 feet above the Colorado River, Amtrak precariously threads its way cliff-side through the long canyon named for Sir George Gore, an Irish baronet who conducted a massive hunting expedition in 1854 in Colorado. It is indeed a twist of fate that Colorado's majestic Gore Mountains, Gore Canyon, and Gore Pass are named for a man whose hunting party was responsible for killing thousands of animals including buffalo, grizzly bear, deer, elk and antelope.

Past the narrow confines of Gore Canyon, the Zephyr enters a more widely sweeping valley of the Colorado River, and hurtles past the hamlets of Radium and State Bridge. At Bond, the Denver and Rio Grande Railroad tracks split with a northerly branch heading to northwestern Colorado's coal mines near Craig. Amtrak stays the course along the Colorado river however, and soon enters the vast expanse of Red Canyon near Burns. From the Zephyr's lounge-dome car the canyon walls climbed forever as the train followed every curve in the river. Then a brief breakout at Dotsero where the Colorado River flows through a wide valley and joins the Eagle River.

Now on the south bank of the Colorado River, the Zephyr plunges into Glenwood Canyon, where man is making a startling transformation of nature. Glenwood Canyon is now filled with sweeping bridges, tunnels, and four lanes of highway as construction of Interstate 70 heads west. The view from the dome car

Heading west into the Tunnel District on April 29, 1989, this view from the train looks toward the rear of the train back to the east.

The California Zephyr heads into another tunnel on its way west to the Moffat Tunnel and Winter Park.

Above left: The Zephyr in Fraser River Canyon between the towns of Fraser and Granby.

Above right: The Zephyr will soon be entering this tunnel heading west in Gore Canyon, above the rushing Colorado River.

The author undertook the road trip above Gore Canyon on May 27, 1989, and photographed Amtrak's California Zephyr heading west, inching along the steep walls of the canyon.

The Zephyr with three engines, three baggage cars, and twelve passenger cars creeps westward through the Gore Canyon tunnel on May 27, 1989.

Above: The train has just about completed its journey through the tunnel.

Below: Having passed through the tunnel, the Zephyr continues its journey west down Gore Canyon.

The Zephyr heads west towards less perilous territory out of Gore Canyon.

An eastbound Zephyr was spotted heading out of Red Canyon and toward Gore Canyon on its journey to Denver on May 28, 1989.

Red Canyon in the vicinity of Burns with the railroad bridge crossing the Colorado River.

is now one of massive cranes, cement trucks and mountain-sized graders and dump trucks.

Glenwood Springs awaits at the west end of the canyon, and we depart the rails for an overnight stay at the Hotel Colorado and a soak in the Hot Springs Pool. Our return trip to Denver the next day is just as spectacular, as the canyons are traversed in reverse order, and Amtrak's comfort and service help us enjoy the relaxing journey all over again. More snow has fallen in the high country, and after a late dinner in the dining car we descend into Denver's western suburbs with city lights twinkling and the ground blanketed with snow.

The historic Glenwood Springs railroad station, built in 1904, is the start of the traveler's journey into the town's history which is as colorful as its canyon walls. Train passengers walk from the station across a bridge over the Colorado River to the Hot Springs and the Hotel Colorado. The hot springs had long been known to the local Ute Indians as a place of healing waters. White settlers arrived in the valley in the 1870s and by 1888 had enclosed the first hot springs pool. A bath house was completed next to the pool in 1890.

Construction of the fabulous Hotel Colorado just north of the hot springs began in 1891. It was to have 200 private rooms and forty private bathrooms. There was also to be a ballroom, billiard room, presidential suite, and private dining room. The hotel opened on June 10, 1893, and still receives guests today. One of its most famous guests was President Theodore Roosevelt who stayed in the hotel in 1905 during a

As the Zephyr entered Glenwood Canyon on April 29, 1989, passengers had a great view of the construction of Interstate 70 through the canyon, including the towering cranes used to place girders on top of piers.

Three levels of finished bridge construction are seen here from the Zephyr's windows. The top bridge carries westbound Interstate 70 traffic, the middle bridge carries eastbound vehicle traffic, and the bottom bridge is for pedestrians and bicycles.

Still viewing Interstate 70 construction from the Zephyr's windows, this is the eastern entrance to a tunnel for westbound traffic.

This is the western exit of the tunnel, with the new eastbound roadway completed below at railroad milepost 349.

Farther west in Glenwood Canyon, this view from the train back to the east reveals Interstate 70 construction on the north bank of the Colorado River, with the railroad tunnel on the south bank of the river.

hunting trip in western Colorado. Rocky Mountain Railroad Club members enjoyed a prime rib dinner in the Teddy Roosevelt Room of the hotel during their overnight stay in 1989.

Amtrak passengers and Interstate 70 drivers today pass by the historic Shoshone Hydroelectric Plant in Glenwood Canyon:

> The Shoshone complex is significant for being one of the earliest hydroelectric plants on the Colorado River and one of the largest in the Rocky Mountain Region to depend upon the flow of a river for its source of power rather than on the stored water of a reservoir. It is also significant as a remarkable engineering accomplishment in terms of the physical difficulties of construction within Glenwood Canyon and the scale of the undertaking.[3]

The hydroelectric complex consists of a diversion dam on the Colorado River 2 miles upstream from the power plant. The dam was constructed from 1907–1910. A 2.5-mile tunnel constructed from 1907–1909 takes water from the dam to two metal penstocks above the power plant. The penstocks are 9 feet in diameter and 287 feet long with a drop in elevation of 165 feet which delivers rushing water to the power plant turbines below. A 153-mile-long power line was constructed in 1907–1909 from the power plant to carry electricity to Denver. It was a remarkable accomplishment for the time period.

Another remarkable accomplishment was the construction of Interstate 70 through Glenwood Canyon:

> A witch's brew of design challenges, daunting physical constraints, and fiercely promoted opposition made the 12.5-mile segment of Interstate 70 through Colorado's Glenwood Canyon a project of three full decades.[4]

This 1901 William Henry Jackson photograph of Glenwood Springs shows the eight-year-old Hotel Colorado on the right bank of the Colorado River, connected to the town by a vehicle/pedestrian bridge. The railroad bridge across the river is the next bridge west. Just below the trees of the hotel, the long expanse of dark water above the river is the Hot Springs Pool. (*Library of Congress, LC-D4-138000*)

Bathers enjoy the Glenwood Hot Springs Pool in this 1907 stereographic view. (*Library of Congress, LC-DIG-Stereo-1s11177*)

The beautifully restored Hotel Colorado still entertains guests today. (*Wikipedia Commons*)

President Theodore Roosevelt on the balcony of the Hotel Colorado tips his hat to the cheering crowd below in 1905. (*H.H. Buckwalter photo, Library of Congress, LC-DIG-ppmsca-36590*)

The Shoshone Hydroelectric Power Plant is seen in this 1978 view, several years before the construction of Interstate 70. U.S. Highway 6 passes in front of the plant. The twin penstock tubes bring water from the tunnel above, down to turbines below in the power plant. (*Library of Congress, HAER COLO.23-GLENS.V.1.4*)

In this view from Amtrak's Zephyr on April 29, 1989, the upper section of an Interstate 70 bridge is completed above the Shoshone Hydroelectric Power Plant, with old Highway 6 below.

This 1978 view to the west in Glenwood Canyon shows the railroad tracks heading for a tunnel on the left, and the U.S. Highway 6 bridge on the right. In the middle is the dam on the Colorado River. Below the dam is a tunnel that takes water 2.5 miles down-canyon to the Shoshone Hydroelectric Power Plant. (*Library of Congress, HAER COLO.23-GLENS.V.1.38*)

As the Rocky Mountain Railroad Club's Amtrak excursion passed through Glenwood Canyon in April 1989, construction of Interstate 70 was still ongoing, and it was remarkable to view the towering bridges and tunnels being constructed. Views from Amtrak's dome-lounge car windows allowed photographs of this remarkable undertaking.

6

Union Pacific 3985,
Cheyenne to Laramie:
June 17, 1989

The Rocky Mountain Railroad Club's excursion train behind Union Pacific's Challenger engine no. 3985 followed the route of the nation's first transcontinental railroad over Wyoming Territory's Sherman Hill:

> Next to winning the Civil War and abolishing slavery, building the first transcontinental railroad, from Omaha, Nebraska to Sacramento, California, was the greatest achievement of the American people in the nineteenth century … The railroad took brains, muscle, and sweat in quantities and scope never before put into a single project.[1]

Man has always battled the elements in Wyoming, and one such obstacle to the construction of the transcontinental railroad was Sherman Hill, elevation 8,247 feet above sea level. Construction of the Union Pacific Railroad west from Omaha across the plains of Nebraska and eastern Colorado and Wyoming proceeded rapidly in 1867. Rails reached Julesburg, Colorado, in March 1867 and headed west to the newly constructed railroad town of Cheyenne, Wyoming, reaching that point on November 13, 1867. Next came Sherman Hill.

> Sherman Hill! The very name is at once suggestive of the Rocky Mountain range and wide Wyoming skies; of boulder-strewn highlands and breath-taking panoramas; of the enterprising cities of Cheyenne and Laramie, strategically anchored near its base on either side; and of the Union Pacific Railroad, its massive motive power and frequent trains … [Sherman Hill] was the greatest obstacle facing the builders of the Union Pacific on their way west.[2]

But Sherman Hill was conquered in April 1868 and rails then headed downhill to the new town of Laramie, reaching there on May 4, 1868. While the Rocky Mountain Railroad Club trip ended in Laramie, the transcontinental railroad did not. It headed ever westward across Wyoming and into Utah to meet the Central Pacific Railroad at Promontory Summit where the golden spike joining the two railroads was driven on May 10, 1869.

> Parts of the Union Pacific and Central Pacific ran through some of the grandest scenery in the world, but the spot where the two were joined together was improbable and undistinguished. No one had ever lived there and shortly after the ceremony no one would ever again. The summit was just over five thousand feet above sea level.[3]

Speaking of grand scenery, "for the westbound tourist [on U.P. trains], the distant white-capped peaks of the Snowy Range rising into view over the summit of Sherman Hill are nothing less than spectacular."[4] Members of the Rocky Mountain Railroad Club were treated to this scenery with seven photo runbys of their westbound train and four runbys on the return eastbound journey. It was a spectacular day on June 17, 1989, from Cheyenne to Laramie, Wyoming, and back.

Sherman Hill (named after famed Civil War Union General William T. Sherman) provides spectacular views for the traveler, but strangely, it is not on the Continental Divide, which is some 175 miles farther west in Wyoming at South Pass, elevation 7,412 feet. It was Sherman Hill though, which necessitated the construction of many railroad facilities, some of which can still be seen today.

> Through the years the demands of the Hill also brought about the erection of extensive facilities at both Cheyenne and Laramie: sizeable shops, roundhouses, turntables and other facilities, including those for fuel, water and servicing.[5]

At Cheyenne in 1989, Railroad Club members could view the Union Pacific engine shop and old roundhouse from the highway viaduct over the rail yards. Of course, the

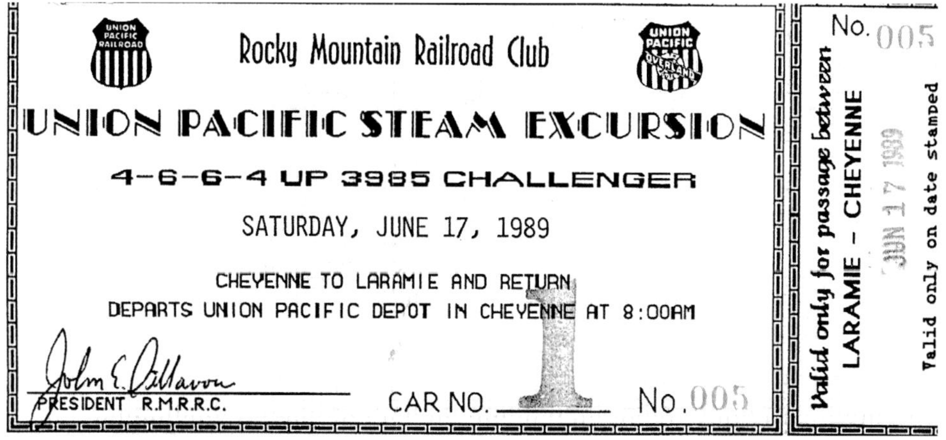

The Rocky Mountain Railroad Club's ticket for the June 17, 1989, excursion behind the Union Pacific's Challenger engine no. 3985.

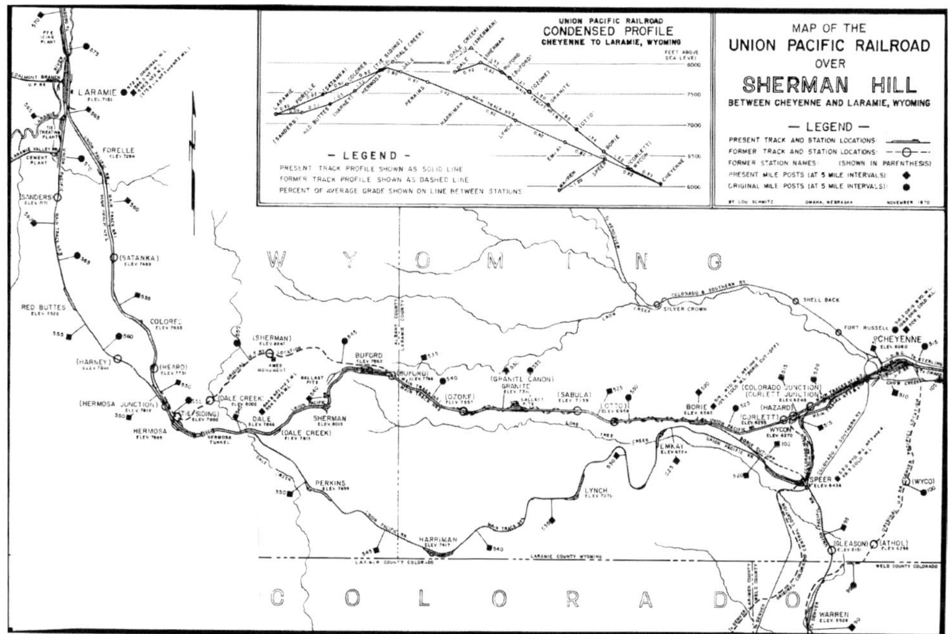

Map of the Union Pacific's rail lines from Cheyenne to Laramie, Wyoming. Tracks 1 and 2—known as the Colores Line (eastbound trains)—are to the north, and Track 3—known as the Harriman Line (westbound trains)—is to the south. (*Rocky Mountain Railroad Club trip brochure map*)

grand 1886 stone passenger depot was on proud display. In Laramie, members viewed the large Union Pacific rail yard from the pedestrian overpass, and the quaint brick depot constructed in 1924 after the original depot burned down in 1917.

Building the rails over Sherman Hill has always been a work in progress. The first transcontinental line went right over the hill at 8,247 feet elevation. From 1880 to 1882, a pyramid monument was built at this spot next to the tracks to honor the Ames brothers, Oliver and Oakes, for their role in the building of the Union Pacific Railroad. Oliver Ames served as president of the Union Pacific from 1866 to 1871, and Oakes was a U.S. representative from Massachusetts. The stone pyramid is 60 feet high and 60 feet wide at the base. On the east side is a bas-relief portrait of Oakes, and on the west side is a portrait of Oliver. This proud monument still stands, however it can only be viewed today from an exit on Interstate 80, as the railroad tracks were moved nearly 3 miles to the south in 1901.

> In 1901 the line over Sherman Hill was relocated to a point approximately two and one-half miles, at the maximum, south of the original line ... The highest point on the old line had been at an elevation of 8,247 feet while the summit on the new line was 8,013 feet, or 234 feet lower than that of the old line. This project required the moving of several million cubic yards of earth for fills and cuts and also involved the construction of the 1800-foot Sherman Tunnel at Hermosa.[6]

The Cheyenne engine house as seen from the highway overpass early in the morning sunlight on June 17, 1989.

Cheyenne station is an impressive site, especially with its huge Union Pacific Railroad sign on the tower.

Rocky Mountain Railroad Club members depart the train at Laramie and head for the pedestrian overpass to view the Union Pacific yards.

The Laramie rail yard with the club's excursion train on the left and freight trains filling the rest of the yard.

Above left: Freight cars in the Laramie yard in September 1941, photographed by Marion Post Wolcott for the Farm Security Administration. Note the water tank and roundhouse at the upper right. (*Library of Congress, LC-USF34-059218-D*)

Above right: Challenger 3985 pulls the club's excursion train through the Laramie yard, with the station on the left.

Jack Boucher photographed the Ames Monument for the Historic American Buildings Survey in 1972. The Union Pacific railroad line had been moved 2.5 miles south of the monument in 1901. (*Library of Congress, HABS WYO, 1-LARAM, 1-4*)

Double-tracking of the main line west from Cheyenne began in 1901 and was completed to the Sherman Tunnel in 1916. In 1918, a second bore was added to the tunnel, allowing double-tracking all the way to Laramie. The tunnels were then renamed the Hermosa Tunnels.

Increased traffic on the Union Pacific required construction of a third line in the 1950s just north of the Colorado border.

Originally it was to replace the old westbound mainline, but for the increased flexibility it provided, the old line was retained. The new line became the westbound line and reduced the maximum grade from 1.55% to .82%, thereby reducing the need for helper locomotives. Completion of this new line took place on May 12, 1953.[7]

Today, the southern westbound line is known at the Harriman Line, named after Edward Harriman, a financier who purchased the Union Pacific in 1897. The double-tracked eastbound line is known as the Colores Line, named after a siding at 7,635 feet elevation located between Hermosa Tunnels and Laramie. Track 3 (Harriman Line) today joins Tracks 1 and 2 at Dale, and they all proceed west through the Hermosa Tunnels until diverging again at Tie Siding.

Pulling the Rocky Mountain Railroad Club excursion train on June 17, 1989, was Union Pacific engine no. 3985, a Challenger class steam locomotive. In September 1936, the first fifteen Challenger type locomotives (4-6-6-4 wheel alignment) were delivered. Eventually, 105 Challengers served on the Union Pacific. The 3985 was retired in 1959 and was stored in the roundhouse in Cheyenne. Restoration of the 3985 was begun in September 1979, and in March 1981, it began operating under its own (coal-burning) power once again. On May 29, 1983, the Rocky Mountain Railroad Club ran the first public excursion behind the 3985 over Sherman Hill. The club's trip in 1989 cost each member $165, and all 270 tickets were sold, resulting in proceeds of $44,500 while the charter trip on the U.P. cost the club $30,000.[8] This was the last trip behind the 3985 before it was converted to oil-burning.

Eleven photo runbys were provided on the Club's 1989 trip: on the Westbound–Harriman Line at mileposts 526.35, 531.5, 537.3, 539.7, 546.0, 551.8, and 553.9; and on the Eastbound–Colores Line at mileposts 560.3, 553.8, 541.3, and 539.1.

Challenger 3985 and tender pose for a photo runby west of Cheyenne.

The club's tail-plate is proudly displayed at the end of the excursion train as passengers re-board after a photo runby.

The Portland Rose was a beautifully restored passenger coach that the author enjoyed on the excursion.

Photo runby no. 1 was at a spectacular landfill at milepost 526.35, near Emkay.

Runby no. 2 featured a head-end shot of the Challenger 3985 at milepost 531.5, east of Lynch.

Club members get ready for runby no. 3 at milepost 537.3, east of Harriman.

The 3985 belches black smoke making for a dramatic runby no. 3.

Runby no. 4 featured a fairly gentle grade for the train at milepost 539.7, east of Harriman.

Club members re-board the train after runby no. 5 at milepost 546.0, west of Harriman.

Above left: The 3985 heads west over the spectacular Dale Creek fill at milepost 551.8 for runby no. 6.

Above right: The Dale Creek fill replaced a series of long spindly wooden and steel trestles over the creek. Swaying in the blustery Wyoming winds from the 1860s to the 1900s, the trestles forced train speeds of no greater than 4 mph.

Runby no. 7 at milepost 553.9 featured cuts through the high cliffs just before the Harriman line merged with the Colores line at Dale. Note the photographers high up on the cliff.

The club's excursion train passes below the cliffs at milepost 553.9.

The 3985 has passed through the gap in the cliffs near milepost 553.9 and heads west to the junction with the Colores line, before both lines head into the Hermosa Tunnels.

After unloading for the lunch stop in Laramie, Club members were treated to an additional "runby" after they had climbed up the pedestrian bridge over the tracks as the train headed west.

The excursion train has passed under the pedestrian bridge and heads west to turn around for the return trip to Cheyenne. The roadway overpass is seen in the distance.

Now turned and heading back east to Cheyenne, 3985 chuffs out smoke at milepost 560.3, south of Forelle on runby no. 8.

A close-up view of runby no. 8.

Climbing out of the Laramie valley, rock formations provide the backdrop for the train at milepost 553.8 near Colores on photo runby no. 9.

The colorful rock formations probably account for the Spanish name Colores at milepost 553.8.

The 3985 chugs around a curve at milepost 541.3 near Buford on photo runby no. 10.

The snow-capped Rocky Mountains are seen in the left background as 3985 cruises by at milepost 541.3.

Spectacular double-trackage is seen in this landfill at milepost 539.1 on runby no. 11 near Ozone.

Above: The 3985 creaks through the double-tracked curve at milepost 539.1.

Right: Club members wave goodbye as the 3985 comes out of the curve at milepost 539.1, heading east to Cheyenne on the final photo runby of the day.

7

Wyoming Colorado Railroad, Laramie to Walden: June 18, 1989

The 92-mile railroad line from Laramie, Wyoming, to Walden, Colorado, had one main purpose—to ship out coal mined near Walden to the Union Pacific railhead at Laramie. While the line from Laramie to Centennial, Wyoming, crosses flat plains, from Centennial the line climbs up the Medicine Bow Mountains through a series of spectacular S-curves, rewarding riders with wonderful vistas. The Rocky Mountain Railroad Club had an excursion on this line on August 4, 1956.

> [The excursion was a] UP standard gauge excursion from Laramie, WY to Northgate, CO over the Coalmont Branch … using [steam] engine #535 [2-8-0 wheel alignment]. The cost was $6.00. Six cars made up the train, all Harriman style coaches with one baggage car.[1]

The club would not make another journey over this branch until after the Union Pacific sold the line in 1987 to the Wyoming Colorado Railroad. The club's excursion was on June 18, 1989, the day following the excursion from Cheyenne to Laramie.

Minerals attracted railroads to this area in the late 1800s, particularly gold and coal. The gold deposits quickly played out, but there were large reserves of coal at Coalmont, 19 miles south of Walden, CO. The first railroad to attempt to build a line into the Medicine Bow Mountains was the Laramie, North Park and Pacific, which operated from 1887 to 1901; it only accomplished building 13 miles of rail west of Laramie. Then on February 27, 1901, the Laramie, Hahns Peak and Pacific Railway was incorporated

> [This used] a connection with the Union Pacific in Laramie … it was to run due west to the tiny mining town of Centennial and up into the heart of the Medicine Bow Mountains to Gold Hill at an elevation above 10,000 feet.[2]

From Walden in Colorado's broad North Park, the line had high hopes of building south to Hahns Peak, CO, to reach the gold mine there, continuing south to Steamboat

The ticket for the club's excursion on the Wyoming Colorado Railroad showed a departure time of 8:30 a.m. from the Laramie station.

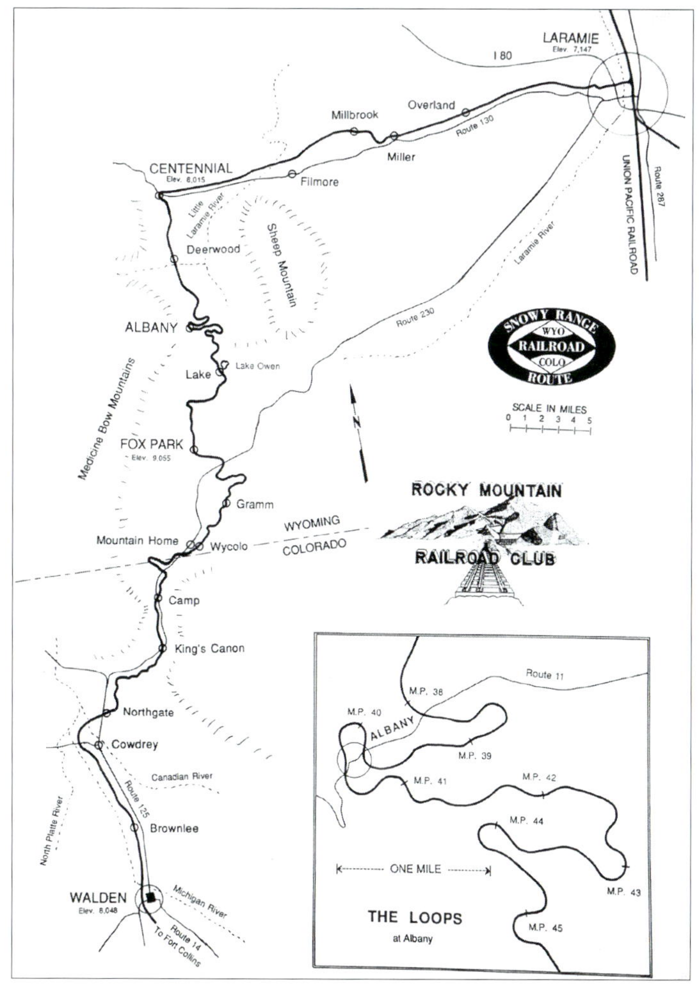

The map of the Wyoming Colorado Railroad depicts the route from Laramie, Wyoming, to Walden, Colorado, with the famous "Albany Loops" 11 miles south of Centennial.

The club's tail-plate
is attached to the rear
of the Appekunny
Mountain observation
car for the trip
over the Wyoming
Colorado Railroad.

Springs and then across northwest Colorado to Utah. This grandiose scheme was never realized.

From 1901 to 1907 the railroad only constructed 30 miles of rail west of Laramie. "It had taken six years to lay just 30 miles of track over practically level ground, one of the slowest pieces of railroad construction in the history of the West!"[3] The 30 miles of rail took the line to the little town of Centennial (elevation 8,015 feet), where an official opening celebration was planned for July 4, 1907.

> The grand celebration on the Fourth of July came off just as planned. Using borrowed U.P. equipment, trains left Laramie at 7:00 a.m., 10:30 a.m., and 2:30 p.m. with the first return trip departing Centennial at 7:30 p.m.… During that one day, the railroad carried 1,200 passengers, a golden spike was driven, and the first section of the Laramie Plains Line was finished and officially in operation.[4]

From Centennial, construction continued south 11 miles to the town of Albany with an elevation gain of 314 feet. Rails reached Albany in November 1908. At Albany, the great loops of the railroad began as it climbed to an elevation of 9,055 feet at Fox Park. This was undoubtedly the most scenic part of the line and it thrilled passengers for nearly 100 years.

WyoColo Railroad heads west across the plains from Laramie to Centennial.

Centennial's railroad station stands across the highway. It was built in 1907 and later relocated from the railroad right-of-way to become a local history museum in 1976.

Just beyond Centennial station, the railroad turns south across the highway to head toward Albany and climb up the Medicine Bow Mountains.

Club members depart the train for the first runby at the Albany Loops.

FP7 engine no. 1512 leads the train for the first runby at the loops.

The train passes through a cut above the lower section of the loop which can be seen in the middle of the photograph.

Appekunny Mountain observation car, formerly from the Great Northern Railway, brings up the rear on the first photo runby in the Albany Loops.

Above left: Club members scramble up the mountain to take advantage of photo runby no. 2 in the upper section of the Albany Loops.

Above right: The Wyoming Colorado Railroad excursion train passes through an upper loop near Albany.

Fox Park was the lunch stop for the excursion. Above the second dome car, the lumber mill facility at Fox Park can be seen.

Club members enjoy a catered picnic lunch at Fox Park, elevation 9,055 feet.

Brandt and Wicklund lumber products were produced in the mill at Fox Park using locally sourced lodgepole pine.

From Fox Park, the rails continued 11 miles south to the Colorado border.

> For the section of the railroad which would run through Colorado, a subsidiary line called the Larimer & Routt County Railway was formed on September 16, 1907 … and the projected route was down Kings Canyon, through Walden, then southwest across North Park to the Northern Colorado Coal Company's property. A company town called Coalmont would be platted at the end of the line. The distance by rail of the Larimer & Routt County Railway was 47.2 miles.[5]

In Colorado, the line descended into Kings Canyon with spectacular views of the snow-capped Colorado Rockies just ahead. "Near the mouth of Kings Canyon the station of Northgate was established and the first train ran to the new station on October 7, 1911."[6] A construction train reached Walden on October 10, 1911, and rail laying continued south to Coalmont. Northgate was the end point for the Rocky Mountain Railroad Club excursion in 1956. The club trip in 1989 went 13 miles farther south to the town of Walden, where club members boarded buses back to Laramie.

The Laramie, Hahns Peak and Pacific Railway reached Coalmont on December 11, 1911. That was the end of the line. No more track was laid. Hahns Peak was never reached. Operating costs were extremely high "due to the severe winters encountered at the high elevation of the line and as a result the line had many reorganizations."[7] The reorganized railroads went by many different names, until finally in 1951, the Union Pacific Railroad bought the line and dubbed it the "Coalmont Branch of the Union Pacific." In 1987, the final owner of the line was the Wyoming Colorado Railroad and its rails were abandoned and pulled up by 2002.

The Rocky Mountain Railroad Club's 1989 excursion on the Wyoming Colorado Railroad had three photo runbys—two at the loops at Albany, and one entering Kings Canyon. The train was pulled by two FP7 diesel engine units nos. 1510 and 1512, which were built new for the Alaska Railroad in 1953. There were seven passenger coaches: four flat-topped cars, which originally had been built for the Southern Railway; two dome cars, which originally served the California Zephyr; and an observation car, the Appekunny Mountain, which was built in 1951 for the Great Northern Railroad. Oddly enough, the Appekunny Mountain was painted in Southern Pacific Daylight orange and black colors, as a result of its participation in a special run from Portland to the World's Fair in New Orleans in 1984. The mash-up of passenger coaches actually resulted in very colorful photos for the members of the Rocky Mountain Railroad Club in 1989.

Engine no. 1512 leads the train through the curves heading out of Kings Canyon on the third photo runby of the excursion.

Colorado's snow-capped Rocky Mountains come into view as the train exits south out of Kings Canyon.

Above left: The gentle grade in North Park signals the train's approach to the small community of Walden and the end of the club's excursion on the Wyoming Colorado Railroad.

Above right: Leading the excursion train was FP7 diesel engine no. 1512, originally built for the Alaska Railroad in 1953. The 1512 and 1510 were leased by the Wyoming Colorado Railroad from the Mountain Diesel Corporation (MDT).

Seven passenger coaches were included on the excursion train seen here at a lower Albany loop. The coaches were four flat-topped cars, two dome cars, and an observation car on the tail.

A conductor stands on the step of one of the ex-California Zephyr dome cars, waiting for club members to return from a photo runby.

The Appekunny Mountain in its bright orange and black Southern Pacific Daylight color scheme brought up the rear of the club's Wyoming Colorado excursion train.

8

Amtrak to the Grand Canyon: May 24–28, 1990

The Rocky Mountain Railroad Club's circuitous excursion to the Grand Canyon over the long Memorial Day weekend of 1990 was a hybrid mixture of tour buses, trains, shuttle buses, and hikes. It all started in Denver with a 5:30 p.m. departure on the evening of Thursday, May 24, with a Grey Lines bus ride to Trinidad, Colorado, some 200 miles south, as there was no longer a passenger rail connection between the two cities. With twilight approaching, the bus passed the twin snow-capped Spanish Peaks 40 miles north of Trinidad.

> The majestic Spanish Peaks, twin mountains rising more than 12,000 feet above sea level in south central Colorado, stand out high above the surrounding Great Plains … The West Peak soars to 13,623 feet, while its companion East Peak rises to 12,708 feet … Surrounding the Spanish Peaks and radiating out in all directions are miles of volcanic dikes … It is estimated that as many as 400 dikes of various lengths radiate out from the Spanish Peaks.[1]

Amtrak's route from Trinidad to Flagstaff, Arizona, followed the rails of the Super Chief, the streamlined passenger train of the Atchison, Topeka and Santa Fe Railway. It was part of the journey from Chicago to Los Angeles which the Santa Fe bragged took only 39½ hours with only one night *en route* westbound. Amtrak took over this route on May 1, 1971, and today, its schedule indicates forty-three hours and ten minutes for the same trip.

The Atchison, Topeka and Santa Fe Railway was chartered in February 1859 to serve the cities of Atchison and Topeka in Kansas and Santa Fe in New Mexico. Its westward rails reached the Kansas–Colorado border in 1873. It hoped to follow the old wagon road of the Santa Fe Trail on to New Mexico's capital city. However, that route meant going over Raton Pass, south of Trinidad. The Denver and Rio Grande Railroad had also pinned its hopes on going over Raton Pass on its planned route from Denver to El Paso and on to Mexico City. Raton Pass "was considered by many

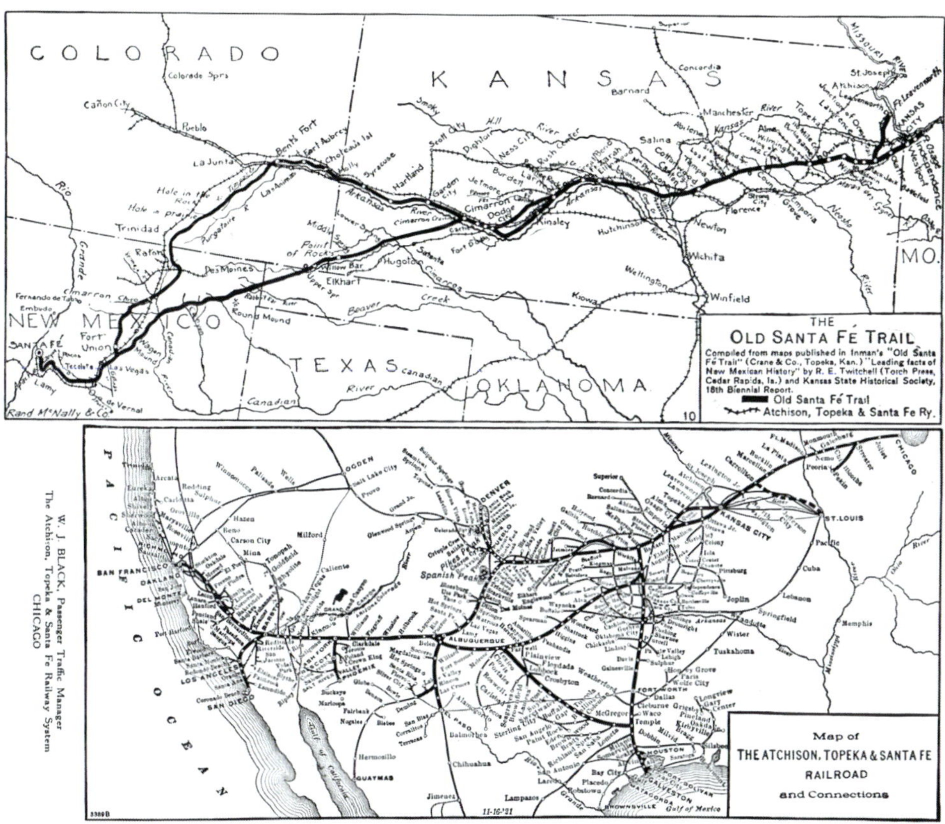

Above: The top map shows how the Atchison, Topeka and Santa Fe Railway followed the old Santa Fe Trail from Trinidad, Colorado, to Santa Fe, New Mexico. The bottom map shows the AT&SF route map from Chicago to Los Angeles. (*Wikipedia Commons*)

Right: Amtrak arrives at Trinidad on May 25, 1990. Club members had to be careful not to trip on the crumbling brick sidewalk when boarding the train.

to be the single most important and strategic mountain pass in the West. It was, at the time, clearly the best way to get a train from Colorado—where the Rockies made construction a nightmare—to California."[2]

> The AT & SF absolutely wanted Raton Pass for its transcontinental railroad. Its surveyors arrived at Raton Pass on the morning of February 29, 1878, and began staking and grading work. Half an hour later, the D & RG surveyors and crew arrived at the site. The competing crews faced each other and prepared for a fight. John A. McMurtrie, the chief engineer for the D & RG, decided to pull his men out and search for another route, an effort that proved unsuccessful.[3]

So the AT&SF won the battle for Raton Pass.

> Its track crews went to work the next day. On 7 December 1878, the first train traveled over Raton Pass by means of a switchback. The next year crews bored a 2,000-foot tunnel through the summit, reducing the maximum grade from 316 feet per mile to 185.[4]

From Raton Pass, the AT&SF descended to the town of Raton, New Mexico. This was the first place in New Mexico where entrepreneur Fred Harvey established one of his lunch counters/dining rooms. Harvey was an immigrant, born in London in 1835. At the age of seventeen, he traveled to New York City and found employment in a restaurant. It was there that he learned the food service business and the need for good food and prompt service. In 1876, he established a food service company to cater to the growing number of rail passengers. Harvey quickly realized that passengers fared poorly for meals on long overland rail journeys west. There were no restaurants to serve passengers at train stops, and no dining cars on the trains. The most a passenger could hope for at a steam railroad's water stop was some rancid meat and cold beans at a roadhouse at the stop. Harvey began to establish dining rooms (and later hotels) at train stops to provide passengers with fresh food and coffee. His dining rooms were so successful that he continued building them all along AT&SF stops on the way west.

At Fred Harvey's dining room in Raton, he hired freed black slaves from the Civil War to serve the diners. This caused hard feeling among the local cowboys, many of whom had fought for the Confederacy during the war. Fistfights and gun battles between the groups soon ensued. The situation became untenable, and Fred fired all the black servers. They were replaced by another radical Fred Harvey idea— white women dressed in prim black and white outfits, who soon became known as "Harvey Girls." "Fred was trying to establish order and civilization in New Mexico during a time that would later be looked upon as a watershed moment for guns and gunslinging in America."[5]

Continuing south from Raton, the AT&SF (Amtrak today) came to the town of Las Vegas, New Mexico. Here, Fred Harvey built the luxurious Castañeda Hotel in 1881. It must have been a welcome sight indeed for the weary, dusty train passengers. The Harvey House hotels and restaurants were built all along the route

Fishers Peak, south of Trinidad, rises above Interstate 25 in this photo taken from Amtrak as it started the climb to Raton Pass.

On May 25, 1990, Amtrak climbs Raton Pass, nearing the tunnel under the pass.

to California. Most have since been destroyed, but a few still stand. The Castañeda was closed in 1948 and sat forlorn and decaying in Las Vegas next to the AT&SF depot until it was restored and reopened in 2019 with great fanfare and tours of the building.

Other Harvey facilities on the route to the Grand Canyon were as follows:

Lamy, NM: El Ortiz Hotel, 1881–1938 (demolished)
Santa Fe, NM: La Fonda Hotel, 1926–present
Albuquerque, NM: Alvarado Hotel, 1902–1968 (demolished 1970)
Gallup, NM: El Navajo lunchroom and dining room, 1923–1957 (demolished)
Winslow, AZ: La Posada Hotel, 1887–1957; restored 1997
Williams, AZ: Fray Marcos Hotel, 1908–1954; restored 1989
Grand Canyon, AZ: El Tovar Hotel, 1905–present.[6]

The AT&SF originally decided not to build into its namesake city of Santa Fe. The grade uphill from Lamy proved to be too steep and curvy for standard-gauge steam engines. Eventually, with diesel engine power, the line was built covering the 17 miles from Lamy to Santa Fe. Amtrak, however, does not go to Santa Fe, so passengers must unload at Lamy and take shuttle services to Santa Fe. From Lamy, Amtrak powers on westward to Albuquerque, where there is a twenty-minute stop, during which the train is washed. By this time in its journey from Chicago, it has picked up a grimy layer of

The old AT&SF station in Las Vegas, New Mexico, appears to be in great shape in this photo from October 2019. Just to the right of the station is the Castañeda Hotel.

The Castañeda on May 25, 1990, was crumbling in decay as it had been closed since 1948.

In October 2019, Allan Affeldt's restoration work on the Castañeda Hotel was just about complete with only some work on the front courtyard remaining.

On October 27, 2019, owner Allan Affeldt led a tour of the restored interior of the Castañeda Hotel, with women dressed in Harvey Girl period uniforms looking on. This was part of the Fred Harvey History Weekend celebration in Santa Fe, Lamy, and Las Vegas for aficionados known as "Fredheads."

A Fred Harvey car and tour bus depart the Castañeda Hotel, *c.* 1920, for a tour of New Mexico's scenic lands. Fred Harvey sent his hotel guests on day trips where the railroad did not go. (*author photo from a display at the Las Vegas train station*)

dirt. During this stop, passengers are free to get off the train and peruse craft items being sold by Native Americans trackside.

From Albuquerque, Amtrak speeds west across the New Mexico desert toward Gallup. The Amtrak conductor on the Rocky Mountain Railroad Club trip announced that speeds reached between 80–90 miles per hour. Heading west into Arizona, the train stops at Winslow, where Allan Affeldt and his wife Tina Mion have lovingly restored La Posada Hotel. They also later restored the Castañeda Hotel in Las Vegas, New Mexico.

Next stop in Arizona is Flagstaff, where passengers intending to go to the Grand Canyon by train must depart. AT & SF built a bypass in 1960–61 around Williams, AZ, the departure point for the Grand Canyon Railway. So passengers shuttle by bus from Flagstaff to Williams.

The AT&SF provided passenger service between Williams and the Grand Canyon from September 17, 1901, until 1968 when automobile traffic to the canyon surged and train ridership decreased:

> The next step, it appeared, would be total abandonment of the line. Yet a few people kept alive the possibility of its restoration, among them historians and environmentalists concerned about overdevelopment along the Canyon's South Rim … Suddenly … all of the pieces of the puzzle started falling into place. In 1987, the project was taken over by Max and Thelma Biegert, two public spirited investors from Phoenix, Arizona. In January 1989, their new corporation, the Grand Canyon Railway, announced plans to resume passenger service by April 1990.[7]

The Biegerts then sped up their timetable to advance the opening day of the Grand Canyon Railway to September 17, 1989, eighty-eight years to the day since the first arrival of an AT&SF train to the Grand Canyon. The Rocky Mountain Railroad Club was fortunate to have a ride on the Grand Canyon Railway less than one year later in May 1990.

Before the club's trip and the opening of the railroad though, a lot of hard work had to be done in reconstructing the line:

> The project from the very first involved bricks and mortar along with rails, spikes and locomotives. Rehabilitation of existing equipment and facilities, and construction of maintenance shops occupied virtually every hour between 29 March and 17 September, 1989 … The Company began rehabilitation of the right-of-way and the Fray Marcos Hotel and depot in Williams.[8]

Thus, coming full-circle, Fred Harvey's Fray Marcos Hotel in Williams has been restored, Harvey's El Tovar Hotel at the Grand Canyon Hotel remains open to this day, and trains run daily between Williams and the Grand Canyon—a story with a happy ending for all.

Forty-four members attended the Rocky Mountain Railroad Club's trip to the Grand Canyon on May 24–28, 1990, so one bus could accommodate the entire group.

Heading west from the dusty plains of Las Vegas, Amtrak climbs the New Mexico mountain ranges toward Glorieta Pass, Lamy, and Santa Fe on May 25, 1990.

Above left: The old AT&SF station at Lamy still stands as seen in this September 2020 view, with Amtrak stopped to drop off passengers for the shuttle ride 17 miles to Santa Fe.

Above right: The Albuquerque transit station serving buses and Amtrak trains is seen here on May 25, 1990. The station was built in the style similar to the nearby Fred Harvey Alvarado Hotel, which was demolished in 1970.

Club members and other passengers enjoyed a chance to stretch their legs at the Albuquerque station as Amtrak paused for twenty minutes to get the dirt washed off the windows.

Above left: The unique washing apparatus was hoisted on a forklift to reach the upper windows of Amtrak's Superliner coaches.

Above right: Club members took advantage of the stop in Albuquerque to peruse the handmade crafts of Native Americans selling their wares next to the train tracks.

In contrast to the Southwest style of the train stations in New Mexico, the Flagstaff, Arizona, station was unique in its Tudor-style architecture.

Fred Harvey operated the Fray Marcos Hotel in Williams, AZ from 1908–1954. The hotel was restored in 1989 by the Grand Canyon Railway and was used for offices and a museum. On May 26, 1990, steam engines nos. 18 and 29 ready for departure from Williams to the Grand Canyon.

Completing the 63.7-mile journey from Williams to the Grand Canyon, the train stops at the old AT&SF station (right), with Fred Harvey's El Tovar Hotel above. The Grand Canyon itself is just beyond the hotel, which sits on the rim of the canyon.

Hundreds of passengers rode the two daily Grand Canyon Railway trains to the canyon in 1990. El Tovar Hotel is seen above the train.

Left: Just beyond El Tovar, the full beauty of the Grand Canyon is on colorful display, with the Bright Angel trail into the canyon below.

Below: It is hard to browse books in the Grand Canyon bookstore with all that scenic beauty just outside the store.

On Friday morning May 25, the bus took members from the Trinidad Holiday Inn for a brief ride through downtown Trinidad's historic district. "Trinidad became a favorite rendezvous for early trappers, traders, and travelers, and was called Santisima Trinidad, 'most holy trinity.'"[9] Trinidad's brick-paved streets and brick buildings certainly did reflect its nineteenth-century heritage. Next stop for the tour was the "Amshack," the derogatory term for the small metal building that served as Amtrak's station in Trinidad. Amtrak's Southwest Chief was due in from points east at 10:40 a.m., but as usual, it was late, arriving at 11:30 a.m. Boarding the train in Trinidad, Club members rode all day and late into the evening arriving at Flagstaff, Arizona at 9:40 p.m., only twenty minutes late. Time had been made up as the train sped at 80–90 miles per hour across the New Mexico plains west of Albuquerque.

Saturday brought a bus trip from Flagstaff some 30 miles west to Williams, Arizona—the boarding point for the Grand Canyon Railway. Departure time for the steam train to the Canyon was 10 a.m., and some 500 passengers clambered aboard the train for the 165-minute journey covering the 63.7 miles from Williams to the Canyon. At the Grand Canyon, club members boarded a bus for an afternoon tour along the South Rim.

After an overnight stay in Tusayan, 8 miles south of the National Park, shuttle buses took some adventurous Club members to the Bright Angel trailhead, where they departed for a three-mile hike into the Canyon to the Three Mile Shelter Cabin. Going down was easy with the 3 miles accomplished in one hour and ten minutes. Climbing back up out of the canyon in the bright sunshine was much more difficult, requiring two hours and twenty minutes. Still, there was plenty of time to catch the 3:30 p.m. train back to Williams. After arriving in Williams, it was back on the bus for the return to the Howard Johnson's hotel in Flagstaff.

Monday morning's Amtrak departure from Flagstaff was scheduled bright and early for 6:50 a.m. There was a bus scheduled to take club members to the Amtrak station at 6:00 a.m. However, by 6:20 a.m., the bus had not shown up and panic set in. The club tour leader called a local taxi company and they sent over two taxis and a van. Through a series of shuttles between the hotel and the station, all club members and luggage were deposited at the station by 6:45 a.m., just seven minutes before the train arrived. That was a close call—it would have been a long way back to Trinidad without the train. Everyone relaxed once aboard the train and enjoyed the journey back east, arriving in Trinidad at 7:40 p.m. From there, it was another long bus ride back to Denver with an arrival at 11:30 p.m. Some club members who did not live too far from downtown Denver actually made it home to their beds by midnight. It had been a glorious, but exhausting trip.

Left: Some club members hiked 3 miles down into the Grand Canyon on May 27, 1990, on the Bright Angel Trail. Dust and mule dung were constant companions.

Below: Two Grand Canyon Railway trains are ready to head back to Williams on May 27, 1990.

Club members wait for departure to Williams at Grand Canyon Station, with the majestic El Tovar Hotel in the background.

Club members enjoy the scenery from Amtrak near Glorieta Pass, New Mexico, on the trip back east to Trinidad, Colorado, on May 28, 1990.

9

Leadville, Colorado & Southern Railroad: July 28, 1990

At 13.7 miles in length, the Rocky Mountain Railroad Club's excursion on the Leadville, Colorado & Southern Railroad was one of the shortest trips covered in this book. The round trip from Leadville, Colorado north to the Climax molybdenum mine at Fremont Pass left the Leadville station at 2 p.m. and returned at 4:45 p.m. The cost was $14.50 for adults and $8.50 for children under twelve. While the total mileage was short, the trip was long on views with Colorado's highest peaks visible to the west—Mt. Massive at 14,421 feet, and the state's highest peak Mt. Elbert at 14,433 feet. As the train crept slowly up the standard-gauge tracks, there was plenty of opportunity to take photographs from the open cars.

For such a short amount of rail mileage, this railroad has one of the most complicated histories in Colorado. It was all started by the mining boom in Leadville, Colorado's highest city at 10,208 feet. There, in the 1860s–70s, gold and silver deposits were discovered in diggings around the booming village. Leadville was "one of the biggest, wildest, richest and longest-running mining camps in the American West. With a flourish and a shout, Leadville put its stamp on a generation of mining men."[1]

At first, horse and mule-drawn wagons hauled the ores over the mountains to Denver, but by 1870, the new railroad companies in Denver set their sights on building rails to the riches of Leadville. The Denver & Rio Grande Railroad had a head start with construction to Colorado Springs in 1871. From there, the railroad headed south to Pueblo, and then west through the Royal Gorge along the Arkansas River, finally reaching Leadville on July 20, 1880. Its competition was the Denver South Park & Pacific Railroad (DSP&P), which was chartered in 1873 with the aim of reaching Leadville and providing competition for the D & RG. "The route selected by the DSP & P meandered back and forth, crossing the Continental Divide twice before reaching Leadville."[2] Both railroads would be constructed as narrow-gauge lines, with 3 feet between the rails, to enable trains to conquer the steep grades and curves of Colorado's mountains. The DSP&P had a much more difficult route to build than the D&RG, which used the broad valley of the Arkansas River to build north from Salida to Leadville.

Afternoon clouds build over Colorado's highest peaks west of Leadville, as the Leadville, Colorado, and Southern train awaits its 2 p.m. departure from the station.

The DSP&P started building southwest from Denver in 1874, following the South Platte River, crossing Kenosha Pass into South Park and then south to Buena Vista, where it joined the D&RG tracks north to Leadville. This arrangement was made possible by a joint operating agreement signed by the two railroads on October 1, 1879. "The D & RG was granted the right to build track from Buena Vista to Leadville, and the DSP & P was granted the right to use the D&RG track to Leadville."[3] The D&RG then completed its tracks into Leadville on July 20, 1880. Disputes between the two railroads were constant, and the DSP&P finally decided to build their own tracks to Leadville over what they called the "High Line." This High Line would run from Como in South Park north over Boreas Pass to Breckenridge and Dillon, and then south over Fremont Pass to Leadville. This route was opened on October 1, 1884. It resulted in a much shorter and faster route to Leadville. The High Line was 151 miles from Denver to Leadville, while the D&RG route via Colorado Springs, Pueblo, Royal Gorge, and Buena Vista was 277 miles.

> Leadville rejoiced when the High Line opened, for the arrival of the South Park [DSP&P] meant competition for the Rio Grande [D&RG]—which would result in lower transportation rates for everyone.[4]

In an ownership twist, the DSP&P had become a branch line of the Union Pacific Railroad in 1880 when U.P. financier Jay Gould purchased it.

> [For the DSP&P] the cost of operating a railroad across Boreas and Fremont Passes was astronomical. The cost of additional helper engines, coupled with the problems of keeping the line open during the frigid winter months also added to a negative balance sheet.[5]

These tremendous costs placed the DSP&P in receivership in May 1888 due to their massive debts. It was reorganized in 1889 and renamed the Denver, Leadville & Gunnison Railroad.

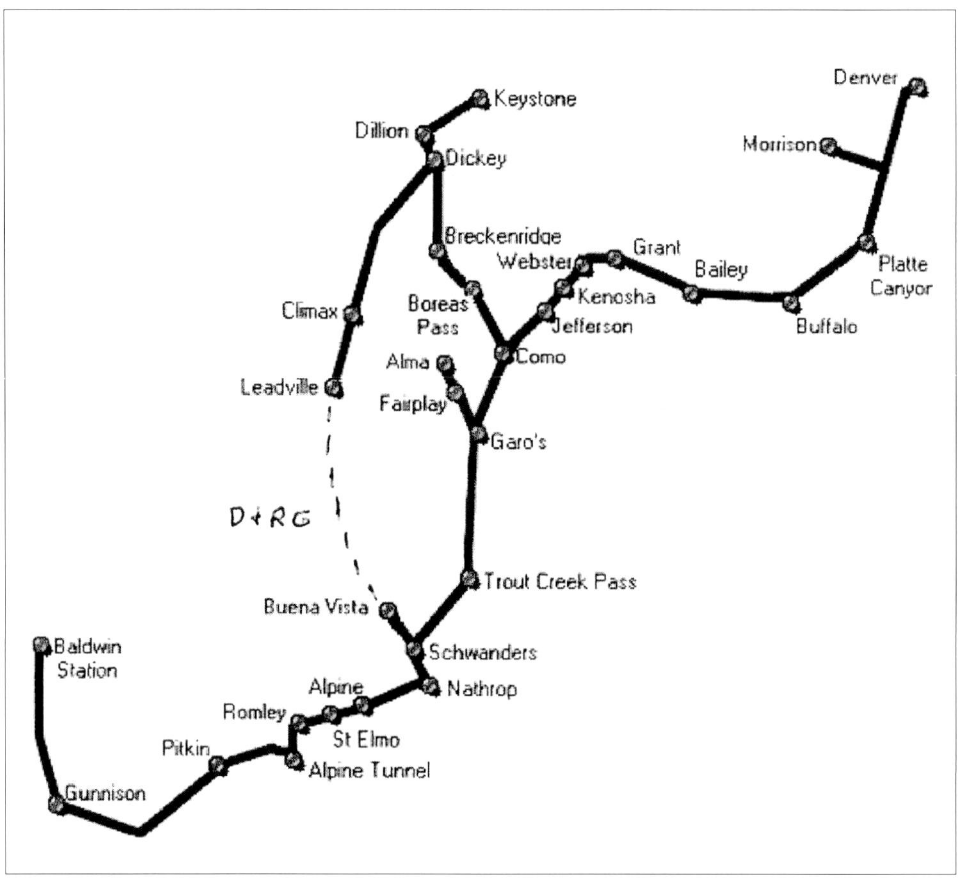

The map illustrates the tortuous route of the Denver, South Park & Pacific Railroad from Denver southwest to Como in South Park, then north over Boreas Pass to Dillon, then south over Fremont Pass at the Climax Mine to Leadville. The extension to Gunnison is shown when the DSP&P was reorganized as the Denver, Leadville & Gunnison Railroad.

> In 1898 the DL & G, still in financial straits, was sold in foreclosure and became one of the narrow gauge routes of the Colorado & Southern Railway ... The C & S operated its narrow gauge branches until the late 1930s when all the routes were dismantled, except for the High Line between Climax and Leadville.[6]

Adding to the railroad name confusion, the Colorado & Southern Railway was sold to the Chicago, Burlington & Quincy Railroad in 1908, however, the High Line retained the Colorado and Southern name. Thus, today's tourist railroad from Leadville to Climax is named the Leadville, Colorado and Southern Railroad.

The molybdenum mine at Climax saved the railroads when the gold and silver mines around Leadville played out. Molybdenum is a metal used in strengthening steel and it became very important in the construction of airplanes. When the Colorado and Southern Railway abandoned its narrow-gauge lines in the 1930s, it kept the High Line from Climax to Leadville because of the profitability of shipping molybdenum.

Leadville, Colorado & Southern engine no. 1978 prepares for the journey up to the Climax Mine for the Rocky Mountain Railroad Club's excursion on July 28, 1990.

A caboose usually brings up the rear of a train, but on this occasion, the LC&S engine pushes the train up the mountain, so the caboose is on the front of the train and the engine is on the rear. Then for the downward journey, the engine is on the front and the caboose on the rear.

Above: Open-air passenger cars allow for many photo opportunities on the trip up the mountain.

Left: Deceptively in the photograph, the engine is still pushing the train up the mountain, although it appears as if the train is heading downhill.

It converted the narrow-gauge High Line to standard gauge in 1943 and continued shipping molybdenum to the D&RG in Leadville and on to Denver. The Colorado and Southern continued operating the High Line until 1986 when the Climax mine was closed. "The High Line route was sold to Stephanie and Kenneth Olsen in December 1987 to operate as a scenic railroad … Colorado's newest railroad opened for business on Memorial Day, 1988."[7] The Rocky Mountain Railroad Club was "johnny-on-the-spot" with its excursion on the line in 1990.

The mountain scenery is certainly the attraction today on the Leadville, Colorado and Southern. Not much remains of the old railway except for the tracks and one water tank at French Gulch, milepost 142.3. This is a 47,500-gallon tank that was originally on the other side of the track from where it is today. It served the narrow-gauge steam engines, and then when the line was converted to standard gauge, the tank was moved to its current location and raised in elevation so that its spout would reach the higher tenders of standard-gauge trains. In 1990, the Leadville, Colorado and Southern made a stop here on the Rocky Mountain Railroad Club's return journey from Climax to Leadville, and passengers could get out, walk around, and take photos. As of this writing in 2022, the line still operates, but like everything else, prices have gone up—$50 for adults and $25 for children. We will always be grateful that we had the opportunity to explore the railroads of Colorado, Wyoming, and New Mexico with the Rocky Mountain Railroad Club from 1987 to 1990.

The peaks near Fremont Pass loom ahead as the Leadville, Colorado and Southern climbs ever upward.

The Climax Molybdenum Mine is an open pit mine at the south edge of Fremont Pass. Late afternoon sun shines on the pit, while the mining buildings are in shadow at the center of the photo.

The train heads back down the mountain, with the caboose now in its normal position at the rear.

Above left: Stopping at the French Gulch water tank, passengers departed the train for a photo runby.

Above right: Passengers have disappeared from view to form a photo line for the passing train.

Back at the station, the Leadville, Colorado & Southern Railroad waits for its next journey up the mountain.

Endnotes

Chapter 1

1 Griswold, P. R., *David Moffat's Denver, Northwestern and Pacific* (Denver, CO: Rocky Mountain Railroad Club, 1995), pp. 53–54.
2 Bollinger, E., *Rails That Climb: A Narrative History of the Moffat Road* (Golden, CO: Colorado Railroad Museum, 1994), dust jacket front flap.
3 Griswold, *op. cit.*, p. 95.
4 Wilkins, T., *Colorado Railroads Chronological Development* (Boulder, CO: Pruett Publishing Co., 1974), p. 220.
5 Butler, M., *Exploring Denver Mountain Parks* (Highlands Ranch, CO: Open Range Publishing, 2012), pp. 125–126.
6 *Ibid.*, p.122.
7 MacGregor, B. and Benson, T., *Portrait of A Silver Lady: The Train They Called the California Zephyr* (Boulder, CO: Pruett Publishing Co., 1977), p. 99.
8 Albi, C. and Forrest, K., *The Moffat Tunnel: A Brief History* (Golden, CO: Colorado Railroad Museum, 1984), p. 10.
9 Patterson, S. and Forrest, K., *The Ski Train* (Golden, CO: Colorado Railroad Museum, 1995), p. 60.
10 Edmonson, H. and Goodheart, D., *Zephyrs Thru the Rockies* (Chicago, IL: Goodheart Publications, 1986), p. 110.

Chapter 2

1 Goss, D., *Journeys to Yesteryear: A Chronological History of the Rocky Mountain Railroad Club* (Denver, CO: Rocky Mountain Railroad Club, 2005), p. 6.
2 *Ibid.*, p. 8.
3 Editors, *History of the Baldwin Locomotive Works 1831–1923* (Milwaukee, WI: Old Line Publishers, reprint 1971), pp. 82–83.
4 cograilway.com

Chapter 3

1 Wilkins, *Colorado Railroads Chronological Development*, p. 67.
2 Bright, W., *Colorado Place Names* (Boulder, CO: Johnson Books, 1993), p. 87.

3 Hayes, W. E., *Rock Island Lines the First Century 1852–1952* (Newton, IA: Circulation Publishing and Marketing, 2000), p. 240.
4 *Ibid.*, p. 245.
5 Rocky Mountain Railroad Club trip brochure, 1987.

Chapter 4

1 Rocky Mountain Railroad Club trip brochure, 1988.
2 Coker, J., 'One More Mountain to Climb' in *Trails Among the Columbine: A Colorado High Country Anthology* (Denver, CO: Sundance Publications Ltd., 1987), pp. 145– 147.
3 Osterwald, D., *Cinders & Smoke: A Mile by Mile Guide for the Durango & Silverton Narrow Gauge Railroad* (Lakewood, CO: Western Guideways, Ltd., 1998), p. 7.
4 Osterwald, D., *Ticket to Toltec: A Mile by Mile Guide for the Cumbres & Toltec Scenic Railroad* (Hugo, CO: Western Guideways, Ltd., 2005), pp. 21–22.
5 Butler, M., *Tracking the Narrow Gauge from Chama to Durango* (Charleston, SC: Fonthill Media and Arcadia Publishing, 2022), p. 56.

Chapter 5

1 Athearn, R., *The Denver and Rio Grande Western Railroad* (Lincoln, NE: University of Nebraska Press, 1977), p. 161.
2 Wilkins, *Colorado Railroads Chronological Development*, p. 61.
3 Gambrill, K. 'The Shoshone Hydroelectric Plant Complex,' *Historic American Engineering Survey* (Washington, DC: National Park Service, 1983), p. 1.
4 Haley, J. L., *Wooing A Harsh Mistress: Glenwood Canyon's Highway Odyssey* (Greeley, CO: Canyon Communications, 1994), dust jacket front flap.

Chapter 6

1 Ambrose, S., *Nothing Like It in the World: The Men Who Built the Transcontinental Railroad 1863–1869* (New York: Simon & Schuster, 2000), p. 17.
2 Ehernberger, J. and Gschwind, F., *Sherman Hill* (Callaway, NE: E & G Publications, 1978), p. 5.
3 Ambrose, *op. cit.*, p. 358.
4 Ehernberger and Gschwind, *op. cit.*, p. 5.
5 *Ibid.*, p. 5.
6 *Ibid.*, p. 29.
7 Rocky Mountain Railroad Club trip brochure, 1989.
8 Goss, *Journeys to Yesteryear*, p. 109.

Chapter 7

1 Goss, *Journeys to Yesteryear*, p. 41.
2 Jessen, K., *Railroads of Northern Colorado* (Boulder, CO: Pruett Publishing Company, 1982), p. 244.
3 *Ibid.*, p. 249.
4 *Ibid.*
5 *Ibid.*, p. 251.
6 *Ibid.*, p. 261.
7 Rocky Mountain Railroad Club trip brochure, 1989.

Chapter 8

1 Butler, M., *Around the Spanish Peaks* (Charleston, SC: Arcadia Publishing. 2012), pp. 7–9.
2 Fried, S., *Appetite for America: Fred Harvey and the Business of Civilizing the Wild West—One Meal at a Time* (New York: Bantam Books, 2011), p. 56.
3 Butler, M., *Tracking the Chili Line Railroad to Santa Fe* (Charleston, SC: Fonthill Media and Arcadia Publishing, 2020), p. 12.
4 Berkman, P. (ed.), *The History of the Atchison, Topeka & Santa Fe* (New York: Smithmark Publishers, 1995), p. 16.
5 Fried, *op. cit.*, p. 72.
6 *Ibid.*, pp. 428–431.
7 Runte, A. *Trains of Discovery: Western Railroads and the National Parks* (Niwot, CO: Roberts Rinehart, Inc., Publishers, 1990), pp. 69–70.
8 Richmond, A. *Rails to the Rim: Milepost Guide to the Grand Canyon Railway* (Williams, AZ: Grand Canyon Railway, 2001), p. 99.
9 Bright, W. *Colorado Place Names*, p. 146.

Chapter 9

1 Blair, E., *Leadville: Colorado's Magic City* (Boulder, CO: Pruett Publishing Company, 1980), dust jacket front flap.
2 Osterwald, D., *High Line to Leadville: A Mile By Mile Guide for the Leadville, Colorado & Southern Railroad* (Lakewood, CO: Western Guideways, 1995), p. 9.
3 *Ibid.*, p. 11.
4 Digerness, D., *The Mineral Belt, Volume II* (Silverton, CO: Sundance Publications, 1980), p. 129.
5 Osterwald, *op. cit.*, p. 16.
6 *Ibid.*, p. 17.
7 *Ibid.*

Bibliography

Albi, C. and Forrest, K., *The Moffat Tunnel: A Brief History* (Golden, CO: Colorado Railroad Museum, 1984)

Ambrose, S., *Nothing Like It In The World: The Men Who Built the Transcontinental Railroad 1863–1869* (New York: Simon & Schuster, 2000)

Athearn, R. *The Denver and Rio Grand Western Railroad* (Lincoln, NE: University of Nebraska Press, 1977)

Berkman, P. (ed.), *The History of the Atchison, Topeka & Santa Fe* (New York: Smithmark Publishers, 1995)

Blair, E. *Leadville: Colorado's Magic City* (Boulder, CO: Pruett Publishing Company, 1980)

Bollinger, E., *Rails That Climb: A Narrative History of the Moffat Road* (Golden, CO: Colorado Railroad Museum, 1994)

Bright, W. *Colorado Place Names* (Boulder, CO: Johnson Books, 1993)

Butler, M. *Around the Spanish Peaks* (Charleston, SC: Arcadia Publishing, 2012)

Butler, M., *Exploring Denver Mountain Parks* (Highlands Ranch, CO: Open Range Publishing, 2012)

Butler, M., *Tracking the Chili Line Railroad to Santa Fe* (Charleston, SC: Fonthill Media and Arcadia Publishing, 2020)

Butler, M., *Tracking the Narrow Gauge from Chama to Durango* (Charleston, SC: Fonthill Media and Arcadia Publishing, 2022)

cograilway.com

Coker, J., "One More Mountain to Climb" in *Trails Among the Columbine: A Colorado High Country Anthology* (Denver, CO: Sundance Publications, 1987)

Digerness, D., *The Mineral Belt, Vol. II* (Silverton, CO: Sundance Publications, 1980)

Editors, *History of the Baldwin Locomotive Works 1831–1923* (Milwaukee, WI: Old Line Publishers, reprint 1971)

Edmondson, H. and Goodheart, D., *Zephyrs Thru the Rockies* (Chicago, IL: Goodheart Publications, 1986)

Ehernberger, J. and Gschwind, F., *Sherman Hill* (Callaway, NE: E & G Publications, 1978)

Fried, S. *Appetite for America: Fred Harvey and the Business of Civilizing the Wild West— One Meal at a Time* (New York: Bantam Books, 2011)

Gambrill, K. "The Shoshone Hydroelectric Plant Complex", in *Historic American Engineering Survey* (Washington, DC: National Park Service, 1983)

Goss, D., *Journeys to Yesteryear: A Chronological History of the Rocky Mountain Railroad Club* (Denver, CO: Rocky Mountain Railroad Club, 2005)

Griswold, P.R., *David Moffat's Denver, Northwestern and Pacific* (Denver, CO: Rocky Mountain Railroad Club, 1995)

Haley, J. L., *Wooing A Harsh Mistress: Glenwood Canyon's Highway Odyssey* (Greeley, CO: Canyon Communications, 1994)

Hayes, W.E., *Rock Island Lines the First Century 1852–1952* (Newton, IA: Circulation Publishing and Marketing, 2000)

Jessen, K., *Railroads of Northern Colorado* (Boulder, CO: Pruett Publishing Company, 1982)

MacGregor, B. and T. Benson, *Portrait of A Silver Lady: The Train They Called the California Zephyr* (Boulder, CO: Pruett Publishing Company, 1977)

Osterwald, D., *Cinders & Smoke: A Mile by Mile Guide for the Durango & Silverton Narrow-Gauge Railroad* (Lakewood, CO: Western Guideways, Ltd., 1998)

Osterwald, D., *High Line to Leadville: A Mile by Mile Guide for the Leadville, Colorado & Southern Railroad* (Lakewood, CO: Western Guideways, Ltd., 1995)

Osterwald, D., *Ticket to Toltec: A Mile by Mile Guide for the Cumbres & Toltec Scenic Railroad* (Hugo, CO: Western Guideways, Ltd., 2005)

Richmond, A., *Rails to the Rim: Milepost Guide to the Grand Canyon Railway* (Williams, AZ: Grand Canyon Railway, 2001)

Rocky Mountain Railroad Club Trip Brochures, 1987, 1988, 1989.

Runte, A., *Trains of Discovery: Western Railroads and the National Parks* (Niwot, CO: Roberts Rinehart, Inc., Publishers, 1990)

Wilkins, T., *Colorado Railroads Chronological Development* (Boulder, CO: Pruett Publishing Company, 1974)